Felicitas
1979

Wolf and Man
EVOLUTION IN PARALLEL

COMMUNICATION AND BEHAVIOR

AN INTERDISCIPLINARY SERIES

Under the Editorship of **Duane M. Rumbaugh**

**Georgia State University and Yerkes Regional
Primate Research Center of Emory University**

DUANE M. RUMBAUGH (ED.), LANGUAGE LEARNING BY A
CHIMPANZEE: THE LANA PROJECT,

ROBERTA L. HALL AND HENRY S. SHARP (EDS.),
WOLF AND MAN: EVOLUTION IN PARALLEL

Wolf and Man

EVOLUTION IN PARALLEL

EDITED BY

Roberta L. Hall

Department of Anthropology
Oregon State University
Corvallis, Oregon

Henry S. Sharp

Department of Sociology and Anthropology
Simon Fraser University
Burnaby, B. C.
Canada

ACADEMIC PRESS   New York   San Francisco   London
A Subsidiary of Harcourt Brace Jovanovich, Publishers

ACADEMIC PRESS, INC.
111 Fifth Avenue, New York, New York 10003

United Kingdom Edition published by
ACADEMIC PRESS, INC. (LONDON) LTD.
24/28 Oval Road, London NW1   7DX

Library of Congress Cataloging in Publication Data

Main entry under title:

Wolf and man.

   (Communication and behavior)
   Includes bibliographies.
   1. Wolves--Evolution. 2. Human evolution.
I. Hall, Roberta L. II. Sharp, Henry S. III. Series.
QL737.C22W64   599'.74442   77-82412
ISBN 0-12-319250-1

PRINTED IN THE UNITED STATES OF AMERICA

Contents

List of Contributors

Numbers in parentheses indicate the pages on which the authors' contributions begin.

MICHAEL W. FOX (19), The Institute for the Study of Animal Problems, Humane Society of the United States, Washington, D.C.

ROBERTA L. HALL (1, 9, 81, 149, 153, 197), Department of Anthropology, Oregon State University, Corvallis, Oregon

FRED H. HARRINGTON (109), Department of Psychology, Mount Saint Vincent University, Halifax, Nova Scotia, Canada

PAMELA McMAHAN (41), 1218 East Roosevelt Road, Long Beach, California

L. DAVID MECH* (109, 133), Endangered Wildlife Research Program, U.S. Fish and Wildlife Service, Patuxent Wildlife Research Center, Laurel, Maryland

ROGER PETERS (95, 133), Department of Psychology, Fort Lewis College, Durango, Colorado

HENRY S. SHARP (1, 9, 55, 81, 149, 197), Department of Sociology and Anthropology, Simon Fraser University, Burnaby, B. C., Canada

MARC STEVENSON (179), Department of Archeology, Simon Fraser University, Burnaby, B. C., Canada

JOHN O. SULLIVAN (31), Department of Biology, Southern Oregon College, Ashland, Oregon

* Present address: North Central Forest Experiment Station, 1992 Folwell Avenue, St. Paul, Minnesota

Preface

This book is like a tree with many roots and branches; the editors have been inspired by many more individuals and intellectual traditions than can be acknowledged here. Our hope is that this work will intrigue readers by presenting significant problems that have too long been ignored or treated with contempt. We promise no easy answers; our goal is to stimulate our readers by drawing their attention to certain crucial attributes of our species, including traits we share with other social predators. If we achieve this goal we shall be more than satisfied.

A major source of our motivation is the treatment given the nature–culture opposition in our society, particularly the absurdity of recognizing a division of the animal kingdom into two categories, human and nonhuman, with the concomitant moral loading of categories. Who has the audacity to analyze human behavior as if it were of a world apart when we can document cultural innovation among Japanese macaques, carry on grammatical conversations with chimpanzees, and report that a wolf would rather answer live wolf howls made by humans than audio recordings of other wolves? Many segments of our civilization, now as in the past, cannot even agree to consider their conspecifics as human if they vary in race, culture, or language. Yet our cultural tradition continues to assert that we as humans are a world unto ourselves, hence denying the animal nature that has permitted our survival and evolution.

A second impetus for this book is provided by recent fossil finds of early hominids, populations of not-quite-human animals that lived from at least 4 million years ago to 1 million years ago. The length of time that these not-quite-humans survived indicates an enviably stable ecological adaptation. If we are to survive as long and as well as even *Australopithecus* we must understand how other animals, with their meager or nonexistent technology, handle the basics of social life.

Empathy and admiration for our own ancestors and their remarkable histories of survival lead us to a third source of inspiration, admiration for the wolf and its fellow gene-pool members, the coyote and the dog. For hundreds of thousands of years wild canids have roamed North America as hunters, developing a rich and successful social life in the process. The arrival of our own species in North America, possibly 100,000 but at least 13,000 years ago, brought a new challenge to the canid population, for the human was nearly as effective a predator as the wolf. Both species fared well under these conditions, as apex predators, sharing a rich environment. Aboriginals admired the wolf and coyote and certainly took no displeasure in their presence, as the remarkably rich traditions of ritual, myth, and medicine among North American native peoples amply illustrate. The canid population found a new econiche and through a new subform, the dog, began a successful exploitation of humanity that continues to this day. These relationships were useful to all involved until the appearance of Europeans with their Christian traditions.

From the beginning, when Genesis commanded man to go forth, be fruitful and multiply, and have dominion over the earth, Western civilizations have grappled with a nature–culture opposition that has allowed no place for nature. The natural world is an affront to Western man; it represents an untamed source of fodder for the maws of his civilization. Our cultural heritage is to subdue nature—to farm it, mine it, log it, and, if all else fails, pave it. Even some of our most ardent conservationists can suggest nothing better than to put boundaries around nature, call it a park, and put it on display like some long-forgotten dust-covered artifact in a museum.

If it is beyond the capacity of our civilization to live with nature itself, how can that civilization coexist with the wild creatures that inhabit nature? Let alone a creature that, like ourselves, is a social hunter of large game, does not hold human boundaries and property laws sacred, and foolishly eats livestock when its wild food is gone—a creature that displays as great a will to survive as does man himself, and brings almost as many resources to the competition.

This animal, the wolf, has been the equal of man since both evolved. Prior to industrialization the wolf had been displaced by humans only under situations of intensive agriculture and then only in limited areas such as the British Isles. Even today, with the full range of industrial civilization arrayed against it, the wolf still holds sway over the margins of the North American continent, those immense areas not suited for agriculture. Only where man practices intensive agriculture and herding can he replace the wolf, and even then only by changing nature itself. In North America, the wolf is man's superior as a hunter, as a predator, and as a species in balance with its environment.

Most of the aboriginal inhabitants of North America recognized these aspects of the wolf and viewed it with respect, affection, or even indifference, but never with fear or hatred. We, as members of Western industrial civilization, are not as sophisticated, for we lay upon the wolf a heavy negative symbolic load. We referred earlier to the *opposition* between nature and culture, but oppositions of this scope contain symbolic statements of lesser generality that are often themselves oppositions; as culture can be represented by man so can nature be represented by animal. Expressed as an opposition between man and animal, some of the symbolic loading becomes obvious, and if we consider the wolf not as an animal but as a specific symbol of animal, then the loading is even clearer.

In our cultural tradition, where nature is intolerable and animals are inferior, the wild wolf becomes not only intolerable and inferior but downright immoral in its refusal to accept the position allocated to it. Only the devil opposes the just and moral goals of Christian Western man, so it is not surprising that the wolf is our prototypical symbol of evil—as demonstrated by an Associated Press news release that reported a wolf on the outskirts of Helsinki; the day after the sighting 500 armed men gathered to pursue it. No other animal is as symbolic of the assumptions upon which our culture is based; no other animal, by its very existence, poses such a threat to our well-being.

Consider the movie *Jaws*. A fictional shark is responsible for a greater destruction of human life than all known wolves in North America and we sell shark T-shirts, patches, and countless other trinkets; what would we do if a wolf were seen on the outskirts of Harrisburg, Pennsylvania?

But the wolf is not the only member of its gene pool in North America, and though our treatment of the coyote parallels our treatment of the wolf, our attitudes toward the dog are mixed. The dog is a mediator between nature and culture; it is an animal but it lives among men as an imitation man. It shares attributes of nature and culture and so shares both symbolic loads. Unlike the wolf, which is seen as unequivocally bad, the dog is seen as either good or bad depending upon which of its sets of attributes is relevant in a given situation.

We shall not issue a direct plea here for the future of the wolf. As anthropologists we are familiar with the operation of symbolic systems and recognize the futility of such a plea. As long as our civilization continues to be based on the fundamental propositions that support it now, it will continue to wreak havoc upon nature and engage in a moral war for the destruction of the wolf. We can only hope that the wolf survives in its margins either until our technology fails or until our system evolves into one that values long-term survival instead of short-term fluorescence.

INTRODUCTION

The Anthropology of the Wolf

In this book we shall examine the wolf—a creature that has long been viewed with fascination and dread by members of Western civilization. We shall consider the wolf and its brother, the coyote, not as objects of Western folklore and mythology but as representatives of a living system that bears many parallels to our own (Figure 1).

What unites wolf and man? One strand in the fabric that binds human to wolf is the dog. Known to be the first animal domesticated by man—an event that happened at least 10,000 years ago, and not just once but at several times, in several places—the dog is the descendant of the wolf. As "man's best friend" it bears many resemblances to the species upon which it has become dependent. Are these similarities due merely to the close association between master and beast? Or are they due to parallel adaptations that occurred when both species were hunters and gatherers, wild carnivores adapting behaviorally to similar ecological zones?

Our position in this book is clearly the second one. We have become intrigued by the similarities between dog and human. We are even more intrigued by the similarities and postulated analogies between wolves and our own australopithecine progenitors, members of a now extinct species.

Certainly the process of "civilizing" the dog has brought about some interesting anatomical changes in this animal. Dogs have slightly shorter snouts and smaller teeth than their wild cousins, and they have developed a suggestion of a forehead. Differences of the same type have been observed in the transition from early to modern human—reduction in the size of the dentition, reduction in the protrusion of the jaws, and en-

1

Figure 1. Human and wolf. Despite the depredations wolves have suffered at the hands of humanity, many modern wolves show a friendly interest in people. These wolves, brought from Alaska's Mount McKinley National Park to the Washington Park Zoo in Portland, Oregon, interact sympathetically with research personnel and zoo visitors. (Photograph by Don Alan Hall.)

largement of the frontal area of the cranium. Though we do not know the significance of these parallel anatomical changes, we can perceive the adaptive value of the highly plastic behavioral response systems that both species possess.

More important for establishing a link between human and wolf than similarities between dog and man are the ecological and behavioral traits that characterize the North American wolf itself. The similarities between the wolf and the hunting–gathering human are impressive. Both are social animals whose living units are relatively small numerically. Both are intelligent, clearly more intelligent than the prey on which they subsist. Both are capable of extreme physical exertion and, more important, of sustained physical exertion at a relatively high level. Both exploit open ground and broken forest areas. Neither species has any single physical attribute that allows it either to overpower or to outrun large prey at will. The two species are about the same body weight and, though not among the largest animals in the world, are virtually immune from predation. Except under unusual conditions, neither can effectively reproduce except as a member of a social unit larger than the conjugal pair.

The demographic structure of the wolf pack provides a good fit with that postulated for bands of early humans. Lone wolves appear to be a rarity, and groups vary in size, ranging up to large and obviously temporary groups of 60 or more animals. "Normal" pack size seems to be about 7 animals. Proportionately, wolves spend the same portion of their lives as infants and subadults as do humans; this may explain in part similarities in the demographic structure and the social structure of wolf packs and hunter–gatherer bands.[1]

Let us briefly review the reasons why the parallel between wolf and hunting and gathering man—which seems so obvious—has not yet received serious consideration.

Following the general acceptance of Darwinian theory 100 years ago, it was agreed that man and ape had stemmed from a common ancestor— that is, shared a common phylogeny or evolutionary history. Though Darwin did not imply that living monkeys and apes were to be regarded as ancestors to humans, this concept was presented by Robert Chambers in his *Vestiges of the Natural History of Creation* (1844/1969) and it was Chambers' simpleminded concept of evolution, rather than Darwin's scholarly one, that became established in the popular mind. In Chambers' view apes were considered near-men, animals similar to ourselves but lacking the important attributes of speech and tool manufacture. Unfortunately, Chambers' view of the evolution of man as a simple progression, an adding-on of complexity that occurs merely because the world was designed that way, still holds the popular imagination.

Among scientists Darwin's view on the critical issue of mechanisms of evolution predominates, with slight changes in emphasis. Darwin postulated ecological adaptation as the key variable that drives evolutionary change. But in the modern view, behavior is seen as the force that initiates the movement of an animal into a new system of relationships with the environment—and anatomy follows behavior. In this interpretation of the mechanisms of evolution, anatomy is not considered determinant. Rather, anatomy adapts in conformity to an animal's relationship with the environment.

---

[1] The rate of maturation of the wolf is absolutely faster than that of man but not really proportionately faster. If we assume a maximum 10-year life span for a wolf in the wild and a 60-year life span for early man, then the relative proportions spent in sexual immaturity are similar. A female wolf can reproduce at age 2 and at 20% of her life span; a male wolf, at 3 years and 30% of his life span; a human female, at 15 to 17 years or approximately 25 to 28% of her life span; a human male, at 17 to 22 years (i.e., at social maturity) and 28 to 36% of his life span. Considering that wolves are seasonal in their reproduction the percentage difference is not terribly great. In both species a very long period is spent in sexual immaturity and it is interesting to speculate whether in fact sexual maturity is a social phenomenon among wolves as well as among people.

How does this view of evolution affect our search for ecological and behavioral analogs to man? Primarily, it makes us question the opinion, inherited from the nineteenth century, that only those animals that bear the closest evolutionary ties to us can illuminate our study of the forces that guided our evolution. Though the great apes—gorillas, chimpanzees, and orangutans—are relatives of man, they are relatives who adapted in a different direction. Their dependence on forests produced behavioral and anatomical traits quite different from those that developed in hominids. Absent in modern apes are three features that are extremely important in the evolution of our own hominid lineage: adaptation to the relatively open savannah country, development of social organization as a requisite for survival, and dependence on social hunting of large game for a small but critical portion of the diet.

The first two traits—living on the savannah and depending on social organization—are found in several groups of monkeys, notably baboons and macaques. Studies of monkey societies were launched in the twentieth century, partly in the hope that models for the ecology and culture of the early hominids would be revealed. Concentration upon these primates brought out some interesting information on their variation and taxonomy as well as on their social life. For instance, it was found that terrestrial monkeys exhibit geographical variability but that speciation—the division of one group into two or more mutually exclusive evolutionary lines—occurs to a greater degree among tree-dwelling primates. This finding has relevance for studies of modern terrestrial humanity, in which geographic variability appears but no tendency to diverge into separate species is found. Wide-ranging ground-dwelling groups of primates tend to inhabit relatively broad niches and to remain relatively generalized.

The study of primate social life and biology has produced useful information pertaining to the hominid line. But because no primate population depends on hunting and meat eating for its day-to-day survival, no primate group can serve as a totally effective model of the life-style of early man. Why is meat eating a critical variable in the life of the hominids? Several popularly read authors—for example, Robert Ardrey—have focused interest on the meat-eating habits of man in the hope that this behavior pattern might provide the clue to modern patterns of war. In fact, hunting and meat-eating behavior has nothing in common with man's aggression against man. However, a hunting system does carry with it significant ecological and behavioral consequences. We shall make four points signifying the importance of hunting:

1. Since meat eaters eat higher on the food chain than do herbivores they must range over relatively more territory in their search for

food. Hence, their density must remain comparatively low and their packs relatively small. In a search for food and for mates they are forced to explore a wider environment.

2. Meat is packaged differently than vegetable food; there is more food value per unit of volume in meat than in vegetables. Hence, less time has to be spent feeding, and more time can be spent searching for food and socializing.

3. Vegetarian primates generally do not share food. The only instances of food sharing observed among primates other than man have occurred when primates—notably chimpanzees—have killed and eaten meat. The important social attribute of sharing, which appears to be the basis of economic reciprocity, clearly appears to be related to hunting as a way of life.

4. Hunting behavior produces a division of labor into two groups, hunters and nonhunters. This division, seen also in wolves, does not follow strictly sexual lines but is also based on age and ability to hunt. Division of labor brings about the important invention of the "home base" where the nonhunters reside and where both groups unite and share their food and their social life. These traits— division of labor and use of the home base—are basic to the lifestyles of some carnivores and of humans but are not as significant among vegetarian primates.

Studies of other animal models could not be pursued intently until the phylogenetic relatives of man had been studied sufficiently to develop and reject as inadequate the "baboon model" of human evolution. Studies of nonprimate species have also been constrained by the opinion that the possession of hands—and hence the ability to manipulate tools—was a primary attribute in the early development of the human niche. However, many instances of various species (for example, the Egyptian vulture and the sea otter) using tools have been documented; yet we know that none of these tool users has adopted a life-style similar to our own. Furthermore, emphasis has shifted from environmental manipulation as a key attribute of human progenitors to ecological adaptation and conceptualizing ability. This is especially true now since we are looking at fossil hominid populations that apparently occupied a stable ecological niche and lived for several million years without patterned stone tools. The effect of these conceptual changes has been to broaden the range of animal models that can be applied to hominid reconstructions.

The opinion that carnivores, and not primates, would provide good models for the adaptation of the hominids has been articulated by George Schaller, who made field studies of the gorilla and of several carnivores in

India and Africa in the 1960s. Observing that studies of primates, though interesting in themselves, would teach little that could be applied to the study of early man, he wrote: "It would also be profitable to observe animals that are phylogenetically unrelated but ecologically similar to the way in which early man probably lived [Schaller, 1973, p. 263]." Behavior, Schaller is convinced, is shaped more by ecological pressures than by family tree. It may also be argued that patterns of variation and speciation tend to be similar in species that adapt to similar ecological zones.

Schaller's main target of study in East Africa has been the lion but he has also observed other major carnivores, including two members of the canid family, the jackal and the hunting dog. In size and social behavior the canids of Africa conform more closely to a model of early man than do the large cats, but it is obvious that no African animal provides an ideal model. Since Africa is considered the major center of hominid evolution it is unlikely that any species would have evolved toward the hominid niche, simply because that niche was filled—and filled well. In the 1950s the common practice was to denigrate the biological capabilities of hominids—man was considered an "unsuccessful ape" or a carnivore without weapons who survived by some accident of fate. Happily, this illogical tendency has been reversed and anthropologists now concede that their ancestors had many positive biological attributes. One finding that has promoted this new position is the growing fossil record indicating that hominids have an antiquity of at least 4 million years, going back into the early Pliocene. And during this time hominids were apparently not particularly rare—their frequency was as great as that of other social carnivores. The record indicates that several forms of hominids were morphologically and ecologically stable over a several-million-year period—scarcely the record of an evolutionary failure!

We cannot expect to find anywhere a biological species filling a niche identical with that of early hominids, with all their unique primate characteristics. But it is possible that on a continent where hominids did not settle until comparatively recently a species of a nonprimate mammalian group may have evolved behavioral and ecological traits similar to those of early hominids. It is argued here that the wolf *Canis lupus* may have been evolving in a direction similar enough to Pliocene hominids to encourage a study of its social organization, culture, and patterns of biological variation that may be used in developing models for early hominid behavior and variation.

Investigation of the wolf as a model for human evolution requires the skills and methodologies of many disciplines. Practitioners of anthropology, psychology, zoology, and ecology have cooperated in producing this book. We hope that this analysis of the wolf will be seen by the reader as holistic, with each chapter supplying a necessary link in the conceptual chain. Yet we can analytically separate several components of the chain and name them: cultural–social patterns, cognition–communication, and paleobiology. These topics will be considered in succession in the book's three parts.

Cultural–Social Patterns. Psychologist Michael Fox discusses conceptual and cultural attributes of human and wolf in Chapter 1. Ecologist John O. Sullivan reports on four aspects of wolf culture in Chapter 2, and Pamela McMahan discusses the natural history of coyote social life in Chapter 3, which presents the coyote as an offshoot of the wolf whose social behavior is influenced strongly by the dominance of human culture. Henry S. Sharp relates the social–ecological adaptation of the wolf to that of one group of human hunters—the Chipewyan—in Chapter 4.

Cognition–Communication. Clearly, cognition in human and cognition in canid have definite points of similarity; Roger Peters discusses these in Chapter 5, the first chapter of Part II. Communication in wolves is the topic covered in the next two chapters. In Chapter 6 Fred H. Harrington and L. David Mech present results of their research on vocal communication; and in Chapter 7 Roger Peters and L. David Mech present the results of studies of scent-marking, a principal communication device used by wolves.

Paleobiology and Classification. Biological variation in wolves and coyotes and in their fossil ancestors is used as a model for discussing the classification and systematics of early hominids by Roberta L. Hall, in Chapter 8, the first chapter of Part III. Marc Stevenson discusses the extinct dire wolf of North America in the same perspective in Chapter 9. The Conclusion presents a discussion of the work completed and sets guidelines for future study.

REFERENCES

Chambers, R. *Vestiges of the natural history of creation.* New York: Humanities Press, 1844. [Reprinted in 1969 with an introduction by Gavin de Beer.]
Schaller, G. *Golden shadows, flying hooves.* New York: Knopf, 1973.

Part I

BEHAVIOR AND CULTURE

Natural selection has long been recognized as a process that results in evolution—a change in the form, structure, or behavior of a species. It is a process that results entirely from natural causes and is devoid of any supernatural or mystical implications, yet at the same time it follows general patterns. Natural selection and its consequence, evolution, are considered predictable to a certain extent. Similar (but not identical) changes are expected to occur in different species that experience similar environmental pressures. For example, the eye in the cephalopod and the eye in the vertebrate evolved independently. Unfortunately, in studies of *human* evolution and the development of culture, recognition of the general nature of the process of natural selection has often been absent. All too frequently, analyses of the evolution of culture have been obscured by unstated philosophical and religious assumptions about man and the nature of the differences between man and animal.

In this book we assume that both culture and evolution are natural phenomena; that culture, therefore, is not necessarily restricted to *Homo sapiens* and antecedent hominids; and that evolutionary pressures similar to those that led to the development of culture in man may be operating on species other than *Homo sapiens.*

The concept of culture, nominally the unifying interest of anthropology, is one that has long troubled anthropologists. It has proved impossible to define culture to the satisfaction of all anthropologists, though attempts continue. It is not our intent to enter this extensive debate directly; instead we shall state what we mean by culture, as the concept is used here, and what, in our opinion, the basic elements of culture are.

The first aspect of our statement about culture is a negative one. Culture is not a moral statement about the difference between man and animal. Though differences of degree in the development of culture exist between man and animal, there is no reason, other than our own self-valuation, to assume that these differences are qualitative. Given the undeveloped state of the analysis of animal behavior, any statements about the nature of differences between man and animal must be regarded as tentative at best, and often merely as species chauvinism.

Minimally, the definition of culture has centered on five aspects:

1. Culture as learned behavior
2. Culture as an adaptive mechanism
3. Culture as a producer of material objects
4. Culture as a symbolic phenomenon transmitted by language
5. Culture as a social phenomenon

We shall examine each of these aspects to determine whether they represent criteria essential to a definition of culture.

Culture as Learned Behavior. The restriction of culture to learned behavior is somewhat arbitrary in that both instinct and learning are organizational and biochemical properties of the brain. These two concepts are imperfectly understood, and in practice it is not possible to differentiate between learned and instinctive behavior. This is true even of complex behavior such as language, which probably involves a genetic predisposition [1] as well as learning. The problem is exemplified by experiments in which chimpanzees have been taught to manipulate symbols of human language and, in effect, to think symbolically. Clearly chimpanzees have not been observed to prac-

---

[1] *Genetic predisposition* refers to the physiological basis for a feature that ethologists call the *fixed action pattern.* This is a behavior that an animal performs at a particular time in its life cycle without outside learning or conditioning. For instance, a cat does not need to be taught to cover its own feces—at a certain age it simply begins to do so. Though ethologists insist that fixed action patterns are innate or instinctive, they generally concede that the patterns become refined as a result of experience—that is, through learning. It is probable that some fixed action patterns underlie human behaviors that we term cultural; yet there *are* differences. Cultural behaviors may have a fixed core, but the range of variability of expression is much greater, and is controlled by tradition and also by individual learning. Thus it is crucial to discover whether variability in social behavior exists between animals of the same species if we wish to determine whether their behavior patterns are cultural.

tice these behaviors in their native habitat—yet it is just as clear that the ability to perform them exists in most chimps.

Insofar as the definition of culture is concerned, the critical aspect of learned behavior is that the learning is supraindividual. An animal must not only be able to modify its own behavioral responses in accordance with its experience; it must also abstract learning from a specific communal heritage to which it can also contribute knowledge it has gained by itself. The task of the investigator is to distinguish between the simple kind of individual learning, of which all animals are capable to some extent, and the complex pattern of learning that we term cultural. To achieve this goal we have been concerned with the flexibility of behavior and the variability of social behavior from group to group within a species. That is, culture as learned behavior should permit the animal group that possesses it to modify existing behavior patterns rapidly.

Culture as an Adaptive Mechanism. Consideration of culture as an adaptive mechanism has been approached in a number of ways. Since each culture has a different capacity to aid the group that possesses it, we can evaluate the "success" of different cultures, either by comparing the number of individuals who share the culture at any given time or by comparing the duration of different cultures. Comparison of cultures is difficult, however, because change is the normal state of affairs. Also, the survival of a given culture in a recognizable form is quite a different question from the survival of the population that bears the culture. Biological descendants of a population have survived long after their native culture was extinguished, and theoretically a culture can be completely transferred from one group to another independently of the biological survival of the original group.

Though the difficulties of comparing cultures from an evolutionary perspective have not been satisfactorily resolved, the problem is not relevant to a study of the earliest stages of the evolution of culture. At that point what is important is that a population's possession of culture—any culture—gives it a selective advantage over cultureless populations vying for the same niche. Generally, anthropologists simply assume that culture per se is an adaptive trait, but since it has been fully described in only one species—*Homo sapiens*—the assumption has no explanatory power and little utility. In attempting to discover and interpret rudimentary culture in other life forms we may be able to explore the conditions under which culture is adaptive.

Culture as a Producer of Material Objects. Anthropologists generally consider the use or production of objects (tools) or the modification of the environment as a characteristic of culture. Although this behavior pattern is unquestionably a part of human culture, it is an unnecessary prerequisite for culture in general. The use of tools by the Egyptian vulture, the sea otter, the chimpanzee, and other animals demonstrates that tool use in itself does not indicate the presence of culture; neither does it indicate that a tool user has a particular mental capacity predisposing it to possess culture. A similar point may be made for a more general statement regarding environmental modification. Many animals alter their environment, yet this trait is not necessarily indicative of culture; it does indicate something about their relationships with their environments. Had our primate predecessors been dominant predators instead of prey their use of material objects would have been quite different.

The relationship between the manipulation of objects, the use of hands, and the development of the brain is of fundamental significance in human evolution but the relationship may be primate-specific; certainly it does not apply to animal families that lack hands, such as the cetaceans. It is our contention that the criterion that culture results in material modification has outlived its usefulness and should be discarded.

The two remaining general attributes of culture are by far the most significant. They are not only prerequisites for culture, but they are also diagnostic features. If these features as we define them are present in the behavioral repertoire of a species, we may feel justified in declaring that the species is a culture-bearing one. The presence of even a single feature is indicative of a transition from a primarily noncultural existence to one that includes, and to some extent is dependent on, culture.

Culture as a Symbolic Phenomenon Transmitted by Language. The symbolic, communicative aspect of culture has produced a great deal of discussion and very little in the way of firm definitions. For our purposes the symbolic aspect of culture relates to the capacity and practice of members of a species to pass information among themselves, regardless of the form in which it is passed. The information, however, must not be merely phatic communion,[2] it must be information about the environment, not just about the emotional state of the communicator. In addition, the information must be passed on by a learned system of communication.

---

[2] *Phatic communion* has been defined as minimal communicative activity, which merely informs listeners that one's channels are open, should it prove useful to say something important. Probably a great deal of human verbal and nonverbal communication falls into this category—no doubt a great deal more than most of us would like to believe.

Regardless of how human languages and their genetic bases are viewed, it is generally recognized that they are the results of an evolutionary process. This implies that languages made a transition from instinctive systems of behavior to learned systems of behavior. At some point the "elements" of the communication system evolved into symbols. We use the word *element* instead of the linguistic terms *phoneme* and *morpheme* for a reason—we do not want to imply that the elements of a communication system are in fact dependent upon frequency, pitch, or any other specific feature of human vocal communication. We wish to allow for the fact that a nonhuman communication system may be based on a process more similar to music than to human language. That is, the elements may not consist of *patterned sounds* but of *patterns of sounds*. The structural relationship between the sounds, not the sounds themselves, conveys meaning. Furthermore, the elements of a learned system of communication need not be vocal but may involve other sensory channels, such as smell, vision, or touch.

In one sense we agree with Leslie White, who contended that the symbol represents one pole of an opposition between learned and unlearned systems. But we must disagree with White's assumption of an opposition between languages and instinctive communication systems. Instead, we believe that a continuum must exist between purely instinctive systems of varying degrees of complexity and capacity, and languages of varying degrees of complexity and capacity. This point has been discussed and defended eloquently in Laughlin and D'Aquili's *Biogenetic Structuralism* (1974), which also discusses neural requirements for language and symboling.

Culture as a Social Phenomenon. Culture is a property of social groups and has an existence beyond that of the individual members of the population that shares the culture. It is a gestalt that shares the three characteristics Emile Durkheim outlined for religion in his classic work, *Elementary Forms of the Religious Life* (1915/1965). Culture in a closed system is general, obligatory, and transmitted from generation to generation. Without denying the uniqueness of the individual and his capacity to introduce change, nevertheless we must submit that culture largely determines the actions and thoughts of individual members of the population.

Before summarizing the foregoing five features, we should briefly discuss the relationship between intelligence and culture. Presumably, as culture develops in complexity it places greater demands upon the "intelligence"—the symbolic capacity—of the population. (The word *intelligence* is not limited to problem-solving ability or the capacity to manipulate objects. It refers instead to speed of response and capacity to learn in new situations as well as to the ability to develop the possibilities of familiar situations.) However, the relationship between culture and intelligence is not clear; what we do know is that high intelligence per se is not a necessary criterion of culture or an indicator of its presence.

The first three features we examined are of differential utility to the analyst. We have discarded the requirement of tool use and environment modification as an unnecessary feature of culture. The definition of culture as an adaptive mechanism is essentially a statement of faith, an *a priori* judgment that culture aids population survival. When a rudimentary form of culture is described in a species other than man it may be possible to examine this statement more closely, and to demonstrate *how* culture aids adaptation and survival. Clearly, this feature does not help to diagnose the presence of a cultural system in another species.

Similarly, the percept that limits culture to learned behavior is basic but too simple; in itself it does not help to diagnose a particular behavioral system as cultural. From an operational point of view the last two features—a population's dependence on learned social behavior systems, which are variable in space and time, and its dependence on learned communication systems—are crucial in diagnosing a particular animal behavior pattern as cultural.

If we adopt these criteria to define culture, we can consider a behavioral system as cultural when at least some of the social patterns of the population can be shown to be learned and flexible; that is, the social patterns vary from group to group. Our reasoning is that a system can be shown to be learned only if there is significant variation in the social behavior of the species; furthermore, this variation cannot come about without a means of communication that is at least partly symbolic.

The set of features we have outlined is basic to any discussion of a system in terms of its being, or capacity to become, cultural. With our present theoretical framework a system that is not learned, is not social, and does not involve meaningful communication among its members could not be considered cultural. As with other kinds of systems, the essence of a cultural system is the subordination of the individual to it. The system exists before and after the life of any individual and it is the system that gives the individual the cognitive equipment with which to organize his life. No matter how unique he is, the individual is born into a supraindividual matrix and must operate within that system with constraints placed on him by his culture. The analyst who wishes to determine whether culture exists within an animal species must first determine whether the constraints placed on an individual are learned ones.

Certainly constraints are placed on the behavior of individuals by noncultural systems. Any statement as to the cultural status of a species is in fact a value judgment and as such is subject to dispute on philosophical grounds. Nevertheless, culture is an evolutionary product and hence must be viewed not as an absolute state but as part of a progression of forms. We have already stated that we think its lower limits lie in the early stages of transition to learned, variable elements of communication and social life.

Part of the difficulty in applying the term *cultural* to the behavior of nonhuman species comes from a deep, historical division in the discipline of anthropology. Does one, after Tylor (1873/1958), regard culture as a separate thing, sufficient unto itself, whose elements are all of equal importance and validity? Or does one, after Morgan (1877/1963), focus on the dynamics of culture and its essential structure?

As holders of the latter position we feel justified in ignoring the presence or absence of many significant aspects of contemporary human culture, such as music and art, and focusing exclusively on two areas, social structure and communication, insofar as these two features program and reprogram the individual's behavior. To emphasize our point, we can contrast the noncultural system of the ant with the cultural system of the chimpanzee. Both behavioral systems depend on social structure and communication; but only the chimpanzee's system, which involves mental and behavioral restructuring due to experience and communication, can qualify for consideration as cultural. We consider the supraindividual feedback network of culture that is woven through and around members of a social unit as basic not only to modern man but to his hominid ancestors. In searching for animal models for the hominid ancestors, we are focusing on the wolf.

The four chapters that follow explore the nature of the behavioral variability of wild canids in North America, and tackle some of the results of the utilization of particular econiches by canids and humans. The lead chapter, by Michael Fox, is an imaginative example of a now traditional approach to comparative ethology. Fox begins with the wolf and seeks insight into wolf behavior by comparing wolf with dog. These two species generate insight into the domestication process, which Fox argues is applicable to human as well as to dog. From this baseline Fox ventures into a comparison of canids and hominids, a search to uncover the biological substratum of human behavior.

This is the traditional approach to comparative ethology: the utilization of our (tentative) understandings of the behavior of other species to shed light upon our own nature as biological and cultural beings.

This approach is useful and, as the great popular interest in the works of Tiger, Fox, Ardrey, Morris, Lorenz, and others indicates, has wide appeal. Illuminating as it may prove to be, however, this approach deals with only half the problem. If the models and concepts developed in the first half of the twentieth century by biologists, ethologists, and psychologists have relevance to the human species, surely the models and concepts developed by social scientists from the behavior of the human animal can be applied to the behavior of other animals. It is in the halfway nature of the approach exemplified by Fox that the anthropologist sees the greatest deficiency in the analysis of the behavior of the canids (and other social animals). As a result of a long tradition of humanism overlaid by a scientific striving for objectivity (justly needed to purge the anecdotal and excessively anthropomorphic nature of earlier explanations of animal behavior), the disciplines most concerned with animal behavior have sought their explanations in behavioristic theories and methodologies.

These approaches, though bringing a much-needed objectivity, have historically been individualistic (egocentric) in nature. As such they are geared to the analysis and observation of individual behavior and seek their explanations at a psychological level. But the wolf shares with man a social nature; both exist, from birth to death, in a sociological matrix. We cannot resist paraphrasing Durkheim's classic injunction that a psychological explanation of a sociological phenomenon is always wrong. An adequate explanation of the behavior of man or animal must take cognizance of sociological as well as psychological phenomena. John O. Sullivan's chapter (Chapter 2) on variability in wolf behavior explores several aspects of wolf behavior and employs sociological as well as ethological and ecological concepts.

Pamela McMahan's chapter (Chapter 3) extends the analysis of the wolf in yet another direction by considering the behavior of the coyote. Wolf and coyote (as well as dog) are interfertile; reproduction occurs across species boundaries in the wild. Though McMahan indicates that genetic differences exist between them (and the dog), it is equally clear that the differences are not a function of genetics alone. Her contribution makes graphically clear the plasticity of behavior of the wild canids in North America and raises additional questions about the biological nature of canids and hominids. If the coyote is only partially social (in part because of human predation), is it safe, considering the full interfertility of wolf, dog, and coyote, to assume that humans are social by genetic disposition alone?

Henry S. Sharp, in Chapter 4, returns to the central issue taken up by Fox but attempts to compare wolf and man from a sociological perspective developed for the analysis of human social behavior. His utilization of a cultural deterministic position, near the outer limits of that approach, illustrates that certain subsistence activities and patterns of dispersal, of marginal interest from an individual perspective, are crucial to the maintenance of both species as systems of social groups. This approach, which is rather new, illustrates the necessity of viewing the behavior of social animals as a system.

By implication, Sharp suggests that wolf social structure clearly indicates the presence of culture. Readers must judge the success of the argument for themselves, but if it is satisfactory then the conclusion is inescapable. The wolf is a culture-bearing species and man is neither as unique nor as alone as Western civilization has thought him to be.

REFERENCES

Durkheim, E. *The elementary forms of the religious life.* New York: Free Press, 1965. [Originally published in 1915.]

Laughlin, C. D., Jr., & D'Aquili, E. *Biogenetic structuralism.* New York: Columbia Univ. Press, 1974.

Morgan, L. H. *Ancient society.* Cleveland: World Publishing, 1963. [Originally published in 1877.]

Tylor, E. B. *Religion in primitive culture.* New York: Harper, 1958. [Originally published in 1873.]

1

Man, Wolf, and Dog

Michael W. Fox

In Western civilizations many tales are told about the wolf, portraying it as a savage killer of defenseless prey. Hunting peoples often are represented as crude savages who kill for food and who mercilessly destroy their own kind if territory is threatened or if basic needs are frustrated. European and American children are told countless fairy tales that depict the wolf as a cunning and conniving monster that will stop at nothing to satisfy its blood-lust. Books such as *The Naked Ape, On Aggression,* and *The Territorial Imperative* (see Ardrey, 1966; Lorenz, 1966; Morris, 1967) purport that humans also are innately aggressive—born killers with instincts that program them to destroy their own kind. These notions are not unlike the folklore and fairy tales about savage wolves that kill and eat children and harmless animals. Freud regarded the unconscious, or the *id,* as the source of man's beastly destructive nature and animalistic desires: Clearly this is a Victorian perception, and one that is neither valid nor useful in describing or managing today's problems. Instead we must take a fresh look at the psychological, social, and physical features that unite human and animal forms. This chapter is such an attempt; it focuses on certain behavioral traits of wolves, dogs, and humans, and the ties that bind them. Analogies between the behavior of wild and domesticated canids (dogs) and that of early hominids and contemporary humans will also be discussed.

Both wolf pack and human tribe practice certain rituals—for instance, ceremonial greeting centered around the leader and gathering and singing before going hunting, after eating, and in the evening (Fox, 1971). These behaviors are social and may also be considered cultural. Whereas the tools of man are acquired, their manufacture and style being passed on from one generation to the next, the tools of the wolf are inherited—sense organs for tracking prey, feet for chasing and digging, and jaws and teeth for catching prey. The wolf's canine teeth are functionally equivalent to the retractile claws of cats, in that the wolf uses them to attach itself to and tear at its prey; by its own body weight and body force it can unbalance and pull down its prey. The wolf's powerful jaw muscles also enable it to disable its prey by evisceration and by crushing muscles so that the prey is soon weakened and unable to run away or offer defense.

These inherited tools, essential for the wolf's survival, may also be used in combat to drive off strange wolves who attempt to join the pack. The population of the pack is kept at a fairly constant level in accordance with the availability and type of prey in a given area. Aggressive behavior thus serves a useful purpose.

Fighting rarely occurs within the pack. Two rivals will display to each other until one, the more subordinate, loses face and withdraws. The dominant or alpha wolf may assert its status by using its jaws to pin the subordinate to the ground—but its jaws do not close. Such is the subtlety of ritual; no one in the pack suffers physical injury, except in rare cases of rivalry over mates or over dominance status. Injury can be avoided by acting submissively toward the top wolf.

In man, we see social organization based on a social dominance hierarchy rather than on physical size, and in modern society on some unique quality of the leader figure (charisma). Similarly, the wolf leader may be physically inferior to others in the pack, but they respect it and rarely challenge it. As it ages, another takes its place, and as in human society the former leader retires to the ranks.

Between-pack and between-tribe communication is facilitated by common biological and cultural rituals. In modern *Homo sapiens* great cultural diversification and varied rituals cause communication breakdowns, yet modern transport and technological interdependence make more frequent social interaction imperative.

Mankind, with its acquired weapons, has to acquire new rituals to prevent its own destruction. The alpha wolf displays its weapons and rarely uses them. Similarly, subordinates respect its weapons, and rarely question its supremacy.

Has humanity, with its new technology, outstripped biological controls of aggression? In the wolf, the subordinate's behavior cuts off the aggression. But in the push-button war, there is no cutoff; both sides are biologically separated by their superior strategic and tactical technologies.

If unleashed, depersonalized, dehumanized war will annihilate us, and we will be buried in the sterile earth of chauvinism. To fight to protect or enforce political or religious ideals or to defend one's national beliefs is the folly of our species. We once fought for land—material territory. We now fight for idealistic territory. In the face of peace and awareness, we may push the button of annihilation, but in the face of a screaming child or submissive foe we might not, for biological cutoff operates. It is easy to kill or maim an enemy when we do not see him.

In both animal and human packs or tribes we see a number of substantially similar phenomena. These include wariness of strangers, even rejection or group attack on unfamiliar individuals; this reaction is termed *xenophobia*. Within the pack or tribe, several rituals keep the group together and regulate social distances between individuals. Group-cohesive rituals or ceremonies in wolf and man include communal singing, displays involving group submission and deference to the leader, and parading by the leader as he displays his status over subordinates (Fox, 1971, 1974).

Another canid pack-hunter, the Cape hunting dog of Africa, has a unique ritual, which—like the ritual of wolves submitting to a non-threatening leader—makes our concept of a dominance hierarchy an oversimplification. This ritual is based on mutual submission, food soliciting, and food giving. One animal solicits and another regurgitates food; a single piece of meat may find its way into several stomachs in this extraordinary ritual. (One of my students, having participated in a Navaho peyote experience, felt that the ritual of oral exchange, with all participants feeling high and later nauseous, had a group-cohesive effect! Clearly the social involvement is of as much or even greater significance to the individual than the mere experience of the drug itself. Similarly the group-coordinated consumption of marijuana or of alcohol has potent socially facilitating effects; an individual therefore may derive a more meaningful experience from the social consequences than from the actual effects of these drugs, which he may not even feel. Socially facilitated group exuberance and involvement overshadows the inebriating effects of the drug, although the latter may facilitate the former.)

Wynne-Edwards (1962), the British ecologist, has described an intriguing social display termed an *epideictic display*, which is seen in vertebrates and invertebrates. The display involves a sudden aggregation of animals, often at a predictable time or place. During the display some kind of (as yet undetermined) information is exchanged between individuals. They may then break up and go their separate ways, or they may remain together as a group and, for example, go to roost, hibernate, or migrate. Both men and wolves show epideictic displays (Fox, 1974). Possibly one individual stimulates another, and there is a kind of magnet effect, so that several individuals are drawn together. Reciprocal stimulation may bring everyone to a similar state of arousal; inhibitions break down; each individual has a similar response threshold and readiness to react. Individuality submerges, and the aggregation reacts like a single social organism with a "collective mind"—a pack, a mob, an army. Animals and humans, when alarmed or threatened by a stimulus that affects all members, respond in a group-coordinated fashion—either panic, escape, and stampede; mass attack; or coordinated defense. Familiar examples in animals are the flight reactions of gazelles, the "mobbing" attacks of small birds that effectively confuse and put to flight such predators as hawks and foxes, and the circular defense that a herd of musk-ox will form when threatened by wolves. Rarely is there time for an epideictic display preceding such reactions; the mere sight, sound, or smell of the alarming stimulus plus communication within the group—alarm signals such as calls or flashing a white tail or belly—trigger the reaction.

Before going hunting, both wolves and Cape hunting dogs have been seen participating in what may well be an epideictic display comparable to the elaborate tribal dances performed before a hunting party sets out. In wolves, howling, yelping, barking, and displays of greeting or active submission (tail wagging and face licking) occur before the pack sets out to hunt (Fox, 1971).

Where does modern man stand in this perspective? We manifest the ancestral patterns and reactions of xenophobia and territoriality, as well as many other group-coordinated behaviors—the hysteria of mobs, filled with real or imagined fears or prejudices; the militant enthusiasm of demonstrators sharing a common cause or grievance; and the unity of spectators vicariously supporting one side in a ritualized form of combat or competition, collectively known as sports (Figure 1), which may trigger extreme violence between contestants or supporters of different teams.

Some researchers are looking for other examples of behavioral abnormalities in natural animal populations, but as yet nothing comparable to the range of disorders in man has been found. This brings us to the intriguing point mentioned in the Introduction, namely, the similarities between the processes and consequences of human culturation and animal domestication. It is only in the domesticated animal, principally in the dog, that we find a range of behavior disorders comparable to those seen in its master and, to some extent, its responsible creator. Most of these disorders, a few of which show a genetic component of familial susceptibility, are essentially a consequence of how the animal is reared and into what kind of social relationship it is raised, a complex including the owners, other members of the family, and other people and dogs in general.

Figure 1. Football game. The wolf pack and human tribes practice certain rituals— ceremonial greeting around the leader and gathering and singing before the hunt. (Drawing by John Slater.)

DOMESTICATION AND CIVILIZATION: SOME ANALOGIES

A phenomenon known as *neoteny* is evident in the structure and behavior of man and dog. Neoteny is the persistence of infantile characteristics, either physical or behavioral, throughout life. The organism essentially remains permanently immature in certain respects. Here we find another analogy between the domestication of animals and the evolution of humankind. *Homo sapiens* and a number of domesticated animals retain immature physical features, such as little hair, relatively thinner skin, a small skeleton or lesser proportion of bone to muscle, and reduced secondary sexual characteristics of males. In dogs, selection has reduced body size and especially the size of the teeth and lower jaw; the genus *Homo* also shows a marked reduction in jaws and teeth.

Neoteny should be distinguished from a prolongation of the period of infancy (Fox, 1976). The latter phenomenon is a feature of the higher primates, including the hominids. The prolongation of infancy affords a greater period of developmental plasticity, during which time the organism literally feeds upon and assimilates a tremendous amount of information. Mason (1968) has emphasized that this phenomenon represents a new evolutionary trend in primates and reaches its climax in man. The prolonged period of infancy enables the subject to acquire information and to elaborate new behavior patterns beyond the constraints of instinctual response capacities. Although the capacity for such development is certainly genetically determined, behavior learned in infancy is more plastic and variable.

Opposite to prolongation of infancy in domesticated animals is the case in which the period of infancy is shortened by special husbandry methods so that growth is accelerated. This is of economic importance, as is the early onset of sexual activity coupled with a lack of mate preference. Dogs may represent an exception, since in some dogs the period of infancy is artificially prolonged, as in the "perpetual puppy syndrome," where the overindulgent owner reinforces the pet's dependent and solicitous behaviors. This pattern may be facilitated through selective breeding to fix neotenous features in toy and miniature breeds such as the poodle, King Charles spaniel, and Chihuahua.

Several of the behavior disorders manifest by these perpetual puppies resemble the attention-seeking neuroses or conversion hysterias of man (Fox, 1968, 1973). These include sympathy lameness and vague abdominal pains, for example, which may be purely psychosomatic in origin, and which are often reinforced or rewarded by the attentions of relatives. These abnormal methods of reducing social distance and gaining attention may appear when the affectional ties are jeopardized by some conflict, such as rivalry and jealousy of a sibling or some other competitor for affection.

On the other hand, social distance tolerance may be severely disrupted by inadequate socialization in which no deep affectional bond is established early in life. Experiments with dogs and monkeys reared in social isolation have shown that such animals are asocial or are fearful and sometimes aggressive if another member of their species comes close to them. It is quite possible that infants reared by cold, affectionless parents, or orphaned children who never develop a close emotional bond early in life, have very different proxemic patterns and may be intolerant of close social interaction.

By virtue of the usual relationship that is established with the owner, the dog closely models the child–parent relationship. Consequently some of the disorders reported in pet dogs are analogous to those appearing in infants. These include conversion hysterias, phobias, anxiety and separation depression, and social maladjustments stemming from inadequate socialization and overpermissive and overindulgent rearing.

Tentatively we may conclude that humanity has brought these behavior disorders upon itself as a consequence of culturation as it has brought them upon the dog as a consequence of domestication. Research on the wolf has given much insight into what domestication has done to its cousin the dog (Fox, 1977). Studies of the wolf are also useful because they increase our understanding of the sociobiological nature and socioecological adaptations of our own hunter–gatherer forefathers.

EARLY MAN–WOLF RELATIONSHIPS

For thousands of years man and wolf have competed for natural resources. The human hunter contributed to the extinction of the saber-toothed tiger, whose niche he absorbed. But *Homo sapiens* was less successful in exterminating the wolf, which also enjoyed the advantages of being a social, cooperative hunter. Instead, hunting peoples sustained a respect for the wolf. Superstition and myth surround this rival hunter even

Figure 2. An intelligent animal. Wolf faces, like this one, often reflect a kind of majesty. (Photograph by Don Alan Hall.)

up to recent times, when government trappers using all their skill and poisons, lures, traps, and high-powered rifles were still outwitted by legendary wolves, such as Bigfoot and Old Three Toes. The Eskimo trappers of Anaktuvik Pass, Alaska greatly respect the wolf, which a human can match only if he has an airplane and telescopic sights and the wolf is caught without cover on a frozen lake.

What makes the wolf such a superior animal? (See Figure 2.) What is the nature of the beast that, up to a point, can withstand the depredations of the modern hunter and at one time shared a similar ecological niche with human hunters? Hunters in the Northern Hemisphere were obliged to live with the wolf, to cohabit, and to share the same prey.

Native domestic dogs were probably hybridized with indigenous wolves in Northern latitudes, and in this way man the hunter indirectly made the wolf his ally.[1] Seeing the great potential that such an ally would offer as a hunting companion, as a guard, or as a draft animal to pull sled or pack, man the hunter sought allegiance with the best hunter of the animal world. With its speed and superior sight, smell, and hearing, the wolf–dog not only improved man's success as hunter but also became, at least in spirit, a part of the hunter, an extension of his own limited sensory and motor abilities, complementing the use of weapons such as spears and arrows.

SOCIAL ORGANIZATION AND ECOLOGY

In *The Emergence of Man* John Pfeiffer (1969) notes that prehistoric hunters had home ranges very similar to those of the wolf but unlike those of any other primate. A wolf pack of 10 may have a home range of 500 to 1000 square miles, whereas a band of about 25 prehistoric hunters had an estimated range of 500 to 1000 square miles. This contrasts with other primate species, whose members never traverse any comparable area during their entire lifetimes. For example, a troop of 40 baboons may use 15 square miles of home range, and one particular troop of 17 gorillas uses a home range of some 15 to 20 square miles.

Prehistoric hunters using so large a home range must have had a great ability to learn their terrain; today, the Australian aborigine knows intimately each tree, mound, and watering hole of his range. This is the aborigine's science, and hundreds of books could be written about the knowledge that these people hand down from one generation to the next.

---

[1] For discussion of the possible origin of the early domesticated dog, see Fox (1977).

Many prehistoric hunters were nomadic, moving through their range and setting up temporary camps, exploiting that area, and moving on.

Some of the similarities between a wolf pack and a group of human hunter–gatherers include an optimal size of the group in relation to the availability of food. In one region a wolf pack might number 12 whereas a group of men might number 15 or 20—these densities tend to be fairly consistent in regions having similar ecological characteristics. In both species there tends to be a social regulation of birth, primarily through rivalry between females in the wolves, and in man through the practice of infanticide and contraceptive taboos. Both wolf and hunting man live well below the full carrying capacity of the environment; that is, the environment could support more individuals than are in fact present. Both wolf and most human groups have a leader or a decision maker and some form of dominance hierarchy or pecking order. Bushmen and aborigine children, and wolf cubs alike, have much freedom and usually are indulged by their elders. Mutual sharing of food exists between young and old, and many rituals control aggression within the group; other rituals function to keep the group together and to reinforce kinship bonds. Pack and group allegiance appears to be maintained through rituals involving dancing and music, and, in the wolves perhaps, through singing and love-in or greeting ceremonies that center on the leader. Where the availability of food varies according to the season, the hunting group and the wolf pack tend to split up, becoming nomadic in one season and more sedentary in another season when food is more abundant.

HUMAN HUNTING COMMUNITIES

I now want to focus on some more specific aspects of the life-style of early human hunters, not in order to make further analogies and tenuous comparisons between man and wolf, but to delve more deeply into our own biological and cultural past, recalling that humans have spent a major portion of their evolutionary history as predators, like wolves. Unlike wolves, we evolved new extractive patterns, including agriculture and a supportive technology, but these developments are extremely recent and much of our basic biological makeup is still composed of programs related to our more ancient and enduring past as hunters.

In Lee and DeVore's *Man the Hunter* Joseph Birdsell (1968) observes that hunter–gatherer populations seem to level off at 60 to 70% of the full carrying capacity of the environment, with social factors stabilizing populations below the absolute level of saturation. Lee and DeVore, in the same book, estimate that the average density of Pleistocene peoples rarely exceeded 1 person per square mile, but normally ranged from 1 to 25 persons per 100 square miles.

Population structure above the family level in hunting–gathering groups also gives some insights into the biology and behavior of early man. Among Australian aborigines, dialect groups or tribes number approximately 500 persons, but these are further subdivided into local groups (bands) of about 25 individuals. According to Birdsell, who has studied Australian societies over a long period, marriage partners are almost always found outside the local group but within the dialect tribe;[2] yet about 14% of marriages in each generation involve spouses from different tribes. Both spatial factors and communication densities are involved in controlling the modal value of the dialect tribe. Birdsell considers that exceptions to the modal value for tribal size result from ecological abundance or from political innovations that allow more intensive communication among local groups.

However, the picture changes when we find humanity increasing its extractive efficiency by means of agriculture, which introduced a new dimension into the ecology as well as into the psychology of the species. A species that until very recently practiced a subsistence economy now is floundering in a technological and essentially domesticated world. We have changed socially but many of the deep-rooted biological and social needs of our past primary mode of existence still linger.

In their contribution to *Man the Hunter,* Washburn and Lancaster (1968) consider that the intellect, interests, basic social life, and emotions of *Homo sapiens* are evolutionary products of an earlier successful hunting way of life. In the last few thousand years before agriculture, hunting became more complex and, in some areas, more sedentary. Dogs were used for pulling sleds; hunting; locating prey, tracking it, and bringing it to bay; and as camp guards. People developed the use of boats in the conquest of water, and used spears and bows and arrows in the conquest of space. Hunters were able to specialize upon large mammals. The impact of human hunters upon large fauna can be indicated by reference to the large parks of East Africa. Where elephants are now protected they are multiplying so rapidly that they are destroying their habitat.

Hunting of large game put a premium on cooperation, planning, and non-aggressive activity; conflict had to be resolved before a hunting party went out. Cooperative hunting by men in groups intensified the sexual division of

---

[2] Following Birdsell's definition (Birdsell, 1968, p. 232), the term *tribe* refers to an aggregation of local groups (bands) in spatial proximity. In Australia, the tribe has a clear identity linguistically and culturally, but lacks political authority.

labor, as females and immature males gathered smaller prey and vegetables. According to Washburn and Lancaster, the way of life of our ancestors was not like that of wolves, and also differed greatly from the life-styles of other primates. As Washburn and Lancaster see it, interest in a large area seems to be an exclusively human trait. However, they do not consider the enormous hunting ranges that wolves also utilize, nor do they consider the fact that wolves, like humans, share food among themselves. Another canid, the Cape hunting dog of Africa, also feeds members of the pack who stay behind to guard the young. Clearly, the canid hunters provide many intriguing social and ecological analogs to early man—more than are provided by any other living animals.

CONCLUDING REMARKS

A study of our past can show us many aspects of our own basic nature and can help us extrapolate from where we have been to where we are now, and perhaps to where we are going in the future. The wolf and other social hunters have helped us comprehend some aspects of our own past as social hunters.

In the wolf pack and in the ancestral hominid hunter–gatherer, we find populations living in a balanced relationship with the ecosystem of the biosphere. As one species—our own—steps part way out of this biosphere and develops an extractive technology, a new ecosystem evolves. This new ecosystem, which is not yet self-contained, not only pollutes the biosphere but depletes it of its natural resources; we exploit the biosphere and we put little back. This technological ecosystem (the *egosphere*) does not yet harmonize with the biosphere, which we are rapidly destroying. The biosphere has become a global ecosystem for the human, who is a megapredator (an ecological misfit, who behaves like a "pioneer" species), as distinct from the wolf, who is a well-integrated apex predator (fitting well as a harmonious "climax" species). Humans are no longer integrated with the biosphere, and in the technosphere that they have developed the system is getting out of hand like a cancerous growth; it is beginning to destroy the human environment as well as to threaten the biosphere. Overconsumption must be reduced along with population growth. If *"Homo technos"* is to develop into true *Homo sapiens* we must develop a closed technosphere that is in equilibrium within itself, and also in an equilibrium relation with the biosphere. If this is not achieved, then the prophets of doomsday will be right. If, instead, both the biosphere and the technosphere are brought into a state of equilibrium, humanity may then be free, in every sense of the word, to develop the third psychic or spiritual level, the ecosystem of the noosphere, as envisaged by Teilhard de Chardin in his book *Man's Place in Nature* (1972).

Instead of alienating humanity from the rest of the organic world, culture must become a bridge, unifying within us both beast and higher awareness (Fox, 1976). New values and self-actualizing needs must be acquired and fostered to replace those materialistic egocentric ones that are contributing to the annihilation of the wolf and all that is wild today, and that tomorrow may destroy all that is human.

REFERENCES

Ardrey, R. *The territorial imperative.* New York: Atheneum, 1966.

Birdsell, J. B. Some predictions for the Pleistocene based on equilibrium systems among recent hunter–gatherers. In R. B. Lee & I. DeVore (Eds.), *Man the hunter.* Chicago: Aldine, 1968. Pp. 229–240.

Fox, M. W. (Ed.). *Abnormal behavior of animals.* Philadelphia: Saunders, 1968.

Fox, M. W. *The behavior of wolves, dogs, and related canids.* New York: Harper, 1971.

Fox, M. W. *Understanding your dog.* New York: Coward, McCann, and Geoghegan, 1973.

Fox, M. W. *Concepts in ethology: Animal and human behavior.* Minneapolis: Univ. of Minnesota Press, 1974.

Fox, M. W. *Between animal and man.* New York: Coward, McCann, and Geoghegan, 1976.

Fox, M. W. *The dog: Behavior and domestication.* New York: Garland Press, 1977.

Lee, R. B., & DeVore, I. *Man the hunter.* Chicago: Aldine, 1968.

Lorenz, K. *On aggression.* London: Methuen, 1966.

Mason, W. A. Scope and potential of primate research. In J. H. Masserman (Ed.), *Animal and human.* New York: Grune and Stratton, 1968. Pp. 101–111.

Morris, D. *The naked ape.* New York: McGraw-Hill, 1967.

Pfeiffer, J. *The emergence of man.* New York: Harper, 1969.

Teilhard de Chardin, P. *Man's place in nature.* New York: Harper, 1972.

Washburn, S. L., & Lancaster, C. S. The evolution of hunting. In R. B. Lee & I. DeVore (Eds.), *Man the hunter.* Chicago: Aldine, 1968. Pp. 293–303.

Wynne-Edwards, V. C. *Animal dispersion in relation to social behaviour.* Edinburgh: Oliver and Boyd, 1962.

2

Variability in the Wolf, a Group Hunter

John O. Sullivan

An apex predator, the wolf demonstrates a superb adaptation, physically, psychologically, and socially. In this chapter I shall discuss four interrelated aspects of that adaptation—group hunting practices, intrapack aggressive behavior, reproduction, and wolf individuality. Though summary statements can be attempted, it is becoming increasingly clear that wolf behavior varies greatly according to place, time, individual personality, and the social and cultural traditions of individual packs. This means that we have much yet to learn from the wolf, and perhaps have many new surprises in store.

GROUP HUNTING

Wolves are social hunters that cooperatively capture prey animals larger than themselves. The ungulates that wolves prey upon range in size from whitetail and mule deer up to moose and musk-ox; clearly the capture of such potentially dangerous prey necessitates caution on the part of the wolves. One of the strategies that has evolved concomitantly with the need for caution and for increased efficiency is a practice called *testing* (Crisler, 1956; Mech, 1966, 1970). Caribou herds north of the Brooks Range in Alaska are tested by the wolves rushing at the caribou. Cripples or stragglers attract the attention of the wolves and are pursued; healthy caribou can outrun the wolves and are generally ignored. Popular notions of hunting strategies such as cutting the Achilles tendon (hamstringing) were not observed by Crisler in the Brooks Range. Rather, the wolves ran full tilt into a caribou, knocked it over, fell upon it, and began to devour it.

The moose, which may weigh up to 900 pounds, is a particularly formidable animal. On Isle Royale, in Lake Superior, moose are tested by wolves before being seriously pursued. If the moose stands its ground, it is not usually attacked; instead, the wolves move on to find easier prey. But if a moose runs, the wolf pack is instantly upon it, tearing at its flanks, and one wolf may attempt to grab the nose in order to slow or stop the moose. Nose biting has also been observed in African wild dogs hunting zebra.

Winter snow depth impedes the progress of large herbivorous mammals, making them vulnerable to wolf predation. Wolves feed on such prey to a great extent through the winter. We have far less information on the hunting behavior of wolves during the summer. In north temperate latitudes, the proliferation of foliage during the summer, in both the canopy and the understory, renders field observation difficult. It is generally believed that wolves prey on small animals, such as rabbits, hares, ground squirrels, and beavers, in the summer. Nonetheless, there is a paucity of data on the hunting of small animals by wolves. Thus, data on the hunting strategies used by wolves in the pursuit of small mammals are of great interest.

Rabbit hunting by European wolves was studied in the national park, Bayerischer Wald, West Germany.[1] Groups of 7 to 11 rabbits were released into wolf-free sections of the $6\frac{1}{2}$ hectare (ha) enclosure and given a few days to learn the location of pathways, feeding areas, brush heaps, and the wolfproof shelter provided. Then wolves were introduced. Individual differences in hunting behaviors and proficiency were noted. Some wolves were diligent hunters, spending a great deal of time either waiting for rabbits to emerge from an exit hole or attempting to flush them from the shelter; these wolves accounted for disproportionate numbers of captures and kills. Other wolves hunted for intermediate amounts of time with moderate success. A few wolves hunted little, and one animal hunted not at all.

One particularly interesting observation was that the dominant male of the pack made no captures, but stood in proximity to the hunt. When a rabbit was caught, he often helped kill it. Hunting strategies such as the head–neck directed bite, as reported in certain carnivores (particularly

---

[1] During the academic year 1974–1975 I studied wolf behavior at the Max Planck Institute in West Germany. Hunting behavior was described in terms of 11 components, and a record of frequency and duration was kept for each animal. If wolves differed in duration and frequency of the components, they were considered to show variability in the hunting of small prey.

felids) were not observed in these tests. Rather, the wolves grabbed the rabbits anywhere they could, by midbody, head, ear, or hind leg. The rabbits were then killed by mouthing. Head shaking, often used by carnivores, was not observed.

Complementary to the observation that the dominant male made no captures was that the lowest-ranked male in the dominance hierarchy was the most successful hunter. This contrasts with the results of other studies that have found a direct correlation between hunting prowess and rank order. Possible reasons for this difference may be *(a)* individual variability in the wolves observed and *(b)* size of the enclosure. Some dominant animals may be enthusiastic hunters, whereas other dominant animals may not be. The same variability could apply at any level in the hierarchy, and without regard to sex. And with regard to size of enclosure, it is likely that dominance relationships are stringently maintained in small enclosures and are more relaxed where the wolves have room to move about. The wolves in the Bavarian forest of West Germany could move freely over some 17 acres. They had the option of participating in the group hunting of rabbits, but they could be hundreds of meters away, totally uninvolved in the hunt.

Wolves in small enclosures may invoke dominance interactions even while the prime activity is hunting, and prey animals show a flight distance when threatened by a predator. It is conceivable that subordinate animals forced into close quarters with dominants will show a similar flight distances. If this tentative behavioral chain operates here, then subordinates that may have the capability for successful hunting behavior could be so intimidated by the proximity of their superiors, with no space available for withdrawal, that their hunting efficiency would be impaired.

Cooperative capture of large animals in a field situation, of course, requires breaching the postulated flight distance. Nonetheless, ample space is available for withdrawal if the subordinate animal is threatened. In close confinement, the animal cannot withdraw, so studies on hunting behavior performed in the usual captive situation might reflect *(a)* the actual hunting prowess of the wolves tested, *(b)* rank order rather than hunting prowess, *(c)* interplay between rank order and hunting prowess, or *(d)* a complex of interacting factors. These factors include total space available, accessibility of withdrawal space, the dominance hierarchy, variability in aggressiveness of the dominant animals and in submissiveness of the subordinate animals in any given wolf pack or in any particular experiment, and the hunting prowess of the animal tested. Clearly, knowledge of these factors or some system of weighting such interacting phenomena is necessary in order to clarify the events occurring during the hunt.

AGGRESSIVE BEHAVIOR WITHIN THE PACK

Seasonal variation has been observed in the incidence of aggression within the wolf pack (Zimen, 1975, 1976). Aggressive behavior is particularly significant in two contexts: *(a)* the regulation of pack size and *(b)* the determination of rank-order relationships in the pack with special reference toward breeding. The general picture of aggressiveness in the wolf pack has been worked out by Zimen through 5 years of close observation of wolves in the wolf preserve of the Bavarian National Forest, West Germany. Zimen found that aggression peaks in the late fall or early winter just prior to the breeding season and continues at high levels throughout the breeding season. Only one female in the pack becomes receptive, and once she comes out of estrus fighting drops in frequency. The appearance of the litter, some 63 days after fertilization, also inhibits aggressive behavior. Wolves do not show aggression toward the puppies, and low levels of aggression continue throughout the summer as the pups develop. In the fall, the battles begin again. This is reminiscent of the fall territorial encounters of the great tit, in which the titmice divide the terrain in such a way that adequate food is available throughout the winter for all territory-holding great tits. It is conceivable that similar forces may be operating through such intrinsic homeostatic regulatory mechanisms to limit wolf numbers. In cases in which the severity of aggressive interaction reaches a high enough level that some wolves are repulsed from the pack, such a mechanism may be postulated.

Pack size varies according to the size of prey taken. Wolf-packs feeding through the winter on small deer may number 3 to 5, whereas packs concentrating on moose may number as many as 15 or 16. Reports of packs of up to 36 wolves may reflect a temporary joining of packs, possibly related animals; they could also reflect the observational method, such as working from aircraft.

As mentioned, the number of agonistic encounters begins to increase in the late fall. At that time, wolves test one another by means of various aggressive displays and through actual fighting (Lorenz, 1954; Schenkel, 1947, 1967). If an individual wolf loses bouts consistently, the other pack members concentrate their attacks on it. Eventually, the loser may be thrown out of the pack, becoming an outcast. Such animals could constitute the trailing wolves reported by Mech, 1970.

A typical encounter in the wolf pack in the Bavarian forest might take place as follows: Finsterau, the dominant female, is in the early stages of estrus. She is followed by the three top-ranked males (Olimook, the dominant male; Nasechen, the beta male, and Gelb Auge, the third-ranked male, who is her 2-year-old son). Nasechen sniffs Finsterau's anogenital region and Olimook steps through the snow until he stands immediately adjacent to Nasechen. Both males face in the same direction. Olimook's tail flies flaglike, high above the horizontal, indicating his dominant position in the pack. His ears are up and pricked slightly forward. He stands tall, like a soldier on parade. His manner gives the human observer an impression of assurance and calm repose; it also is imposing and intimidating. Nasechen, a fat, large, older wolf, swings his scarred muzzle toward Olimook. His tail hangs vertically straight down, ears are pressed flat against his head, and he performs the vertical lip raise. His lips are pulled back into a "smile" and he vocalizes, snarling (Figure 1). Olimook joins in the vocalization; deep, rumbling growls seem to roll up from his midbody.

The two male wolves maintain these postures and vocalizations for a few seconds or minutes. Then, Nasechen either withdraws or slowly be-

Figure 1. Alpha wolf and beta wolf. The alpha wolf's manner gives the observer an impression of assurance and calm repose. The beta wolf, ears pressed flat against his head, performs the vertical lip raise. (Drawing by John Slater.)

gins to fall or roll over on his side, perhaps even to his back. Nasechen's snarl turns to a whimper. Olimook then steps on Nasechen or simply looks down toward his rival, growling with his canines exposed. Olimook's lips are pushed forward in the aggressive posture of the canids. Then the dominant animal steps back, ending the encounter. Finsterau may then move off, closely accompanied by Olimook. After a few seconds, Nasechen scrambles to his feet and follows the dominant animals. The whole process may be repeated a short time afterward.

As ovulation approaches, the encounters intensify. Muzzle biting is now added to the previously mentioned behaviors. Nasechen received many muzzle bites during the course of the breeding season. These are nonlethal bits, but they do penetrate the skin and cause some bleeding, so at the end of the breeding season, Nasechen's muzzle was covered with fresh punctures, scabs, and old scars. Other lower-ranking males (from Gelb Auge, the third-ranking wolf, on down) tended to avoid encounters with Olimook during this time. A steady gaze in their direction from the dominant animal sufficed to keep them a few meters from him and, generally, from Finsterau, since he maintained close contact with her.

To summarize, fighting within the pack serves the important functions of limiting pack size so that the wolves do not exceed the carrying capacity of their environment and determining rank-order relationships, in particular with reference to breeding. Rarely do aggressive encounters result in serious injury, though a few deaths have been reported from the field.

REPRODUCTION

Conflicting reports on the nature of the mating bond of wolves exist in the literature, but monogamy has been suggested as the norm. Machismo duels among the adult males in the pack have been presumed to determine which male pairs and mates with the dominant and receptive female, but in reality this conclusion may not hold at all. The wolf pack in the Brookfield Zoo was dominated by a male that expended so much time and energy on dominating the pack and marking the territory that he appeared to have little of either left for reproductive activities. In this pack the beta male copulated with the receptive female.

In the wolf pack I observed in the Bavarian forest, only Finsterau, the dominant female, was attractive to the males. While she was in estrus (in February) she was closely followed by the five adult male members of the pack. Finsterau signaled her condition by elevating her tail and exposing the swollen vulva. The males sniffed at her anogenital region frequently during this time. Nasechen, the beta male, was the first to copulate with Finsterau.

The copulatory position of the wolf is similar to that of the domestic dog. The receptive bitch stands for the male, who mounts dorsoventrally from the rear. As in the dog, a copulatory tie is achieved, and it lasts approximately 20 minutes. After tying, the male characteristically dismounts and turns so that he and the female face opposite directions with their rumps together. While in this position, Nasechen was often harassed by Olimook, and Nasechen sometimes rolled over to submit while still tied to Finsterau. This must have caused some pain, because Finsterau whimpered and sometimes even turned to snap at and bite Nasechen. Nasechen copulated with Finsterau about half a dozen times over a period of $2\frac{1}{2}$ days. Next, Finsterau copulated with Gelb Auge. After several copulations with this third-ranked male, Finsterau paired off with the dominant Olimook. He bred Finsterau about 10 times and one of these matings resulted in fertilization (determined by counting back 63 days, the gestation period, from the littering date).

In this case the dominant male sired the litter, a feature that has been interpreted as an adaptation for passing on the most adaptive traits. Similarly, in the baboon society observed by Washburn and DeVore, the receptive females presented first to subdominant males and at the height of estrus (the time of ovulation) formed a consort pair with one of the dominant males. This procedure seems to serve not only to pass on the genes of the best-adapted male, but also to reduce strife, both in baboon troop and in wolf pack.

Mating systems of wolves therefore include *(a)* breeding by the monogamous dominant pair, *(b)* the beta male mating with the dominant female while the alpha male maintains the territory, and *(c)* the dominant female copulating with several of the top-ranked males. We can best resolve these conflicting reports by assuming that wolves have a wide range of options available in mating. In addition, individual variability of wolves would be expected to influence the mating system of a given pack. An individual female, for instance, might be highly motivated sexually, although not dominant in most other social situations. On the other hand, the male leader of the pack might lack a strong sexual orientation, or the dominant female might be more strongly sexually or socially oriented than the dominant male. The mating system of a particular wolf during a given breeding season is therefore the result of a number of variables, including the social and sexual orientations of the high-ranking adults of the pack.

Burrow Construction

Wolves dig burrows, several meters in length, in which they litter; they have also been known to enlarge old fox burrows. The pups remain in the safety of the burrow for about 3 weeks, at which time they may venture forth for short distances from the mouth of the burrow. How the burrow is constructed, like other aspects of wolf behavior, appears flexible. The dominant male of the pack at the Biosocial Facility in Eugene, Oregon, assisted the female in the construction of the burrow.[2] However, in the spring of 1975 Finsterau constructed the entire burrow herself, choosing a site different from the one she had been observed to use in 1974. She selected a site at the base of a stump and began digging down through the snow; she dug with her forepaws-and used her nose as a shovel to move the dirt. Soon she began throwing dirt out of a tunnel which was increasing in size. From studies made by Ginny Ryan it appears that the burrow leads back to a slightly enlarged chamber where the pups are born.

INDIVIDUALITY OF WOLVES

Group hunting results in conditions that could favor selection for individual variability, or the division of labor. A wolf pack must necessarily possess a variety of skills in order to match the environmental exigencies presented to it. The pack must have the necessary skills to cope with a spectrum of prey species that vary temporally and spatially in their abundance. Furthermore, each prey species would be expected to have at its disposal a variety of methods for avoiding predators. These would vary from crypticity to fleetness of foot to antipredatory defensive fighting. The wolf shows a great deal of polymorphism in terms of size, pelage coloration, and body configuration. Such physical variability suggests the likelihood of polyethism.

Further evidence for polymorphism in the wolf comes from the American Kennel Club. Over 125 breeds of dogs have been produced by artificial selection from the ancestral gene pool of the wolf. These breeds differ not only in structure but also in behavior. It is therefore reasonable to expect wolves to show individual behavioral differences. Certainly, differences emerged in the rabbit-hunting behavior of the Bavarian pack. Individual wolves, then, differ widely in behavioral repertoires from fellow pack members.

The behavioral repertoire of the pack leader is particularly important, since the overall behavior of the pack is expected to reflect the repertoire

[2] John Fentress and Ginny Ryan provided this information from their studies of captive wolves at the University of Oregon.

of its leader. Comparative work on different wolf packs is sorely needed to clarify this point. Workers studying Japanese macaques have reported differences in such behaviors as awakening time and movements during the reigns of two different troop leaders, Titan and Jupiter. The same principle may well apply to wolves.

If group hunting through natural selection has given rise to individual variability among wolves and, consequently, among wolf packs, the task of describing the behavior and ecology of the wolf becomes exceedingly difficult. The different and sometimes conflicting results reported in the literature could reflect

1. Individual differences among wolves and wolf packs
2. The specific conditions under which observations were made including how the fieldwork was conducted (e.g., from aircraft or by means of radio telemetry) and whether the animals were in close confinement, as found in most zoos, laboratories, and experiment stations, or in large preserves that attempt to simulate the natural habitat but still allow close observation of individual animals
3. The interplay between item 1 and item 2—that is, the individuality of the animals and the effects of the conditions under which they were observed

The wolf, large wild dog of the boreal forests, remains an elusive subject for study. It is to be hoped that this creature will be preserved for future study, or at least that a great deal more information may be gathered before this large, wild predatory canine disappears over the horizon, as have so many of its brethren.

REFERENCES

Crisler, L. Observations of wolves hunting caribou. *Journal of Mammalogy,* 1956, *37,* 337–346.

Fox, M. W. *The behavior of wolves, dogs, and related canids.* New York: Harper, and Row, 1971.

Fox, M. W. (Ed.). *The wild canids.* Princeton, New Jersey: Van Nostrand-Reinhold, 1975.

Lorenz, K. *Man meets dog.* London: Methuen, 1954.

Malcolm, J. R., & Lawick, H. van. Notes on wild dogs *(Lycaon pictus)* hunting zebras. *Mammalia,* 1975, *39,* 231–240.

Mech, L. D. *The wolves of Isle Royale.* Fauna of the National Parks of the United States, Fauna Series 7, 1966.

Mech, L. D. *The wolf: The ecology and behavior of an endangered species.* Garden City, New York: Natural History Press, 1970.

Murie, A. *The wolves of Mount McKinley.* National Park Service, Fauna Series 5. Washington, D.C.: U.S. Government Printing Office, 1944.

Rutter, R. J., & Pimlott, D. H. *The world of the wolf.* Philadelphia: Lippincott, 1968.

Schenkel, R. Ausdruckstudien an wolfen. *Behaviour,* 1947, *1,* 81–129.

Schenkel, R. Submission: Its features and functions in the wolf and the dog. *American Zoologist,* 1967, *7,* 319–330.

Scott, J. P., & Fuller, J. L. *Genetics and the social behavior of the dog.* Chicago: Univ. of Chicago Press, 1965.

Zimen, E. Social dynamics of the wolf pack. In M. W. Fox (Ed.), *The wild canids.* Princeton, New Jersey: Van Nostrand-Reinhold, 1975.

Zimen, E. On the regulation of pack size in wolves. *Zriyschraft für Tierpsychologie,* 1976, *40,* 300–341.

3

Natural History of the Coyote

Pamela McMahan

Although American and Mexican Indians venerated coyotes in their art and legends, the first published accounts of these "little wolves" were written by explorers, conquerors, and visiting naturalists. Mingling descriptions of wolves, foxes, and coyotes, the early journalists spread colorful, but often confusing, tales of the coyote. Lewis and Clark mentioned coyotes in their chronicles, but it was Thomas Say who in 1823 provided the "prairie wolf" with its scientific name, *Canis latrans*.

By the middle of the nineteenth century, the coyote was already marked as an enemy by humans. Superbly capable and adaptable, the small predator had taken advantage of its new food source: livestock. Ranchers and hunters launched a campaign aimed at eliminating coyotes, along with all other predators whose habits conflicted with human interests. With traps, guns, poisons, and other weapons, North Americans battled against the survival of coyotes. Even national parks adopted policies aimed at protecting "game proper" and eliminating predators.

In conjunction with this onslaught, writings about coyotes began to reflect economic interests. Supplemented by reports of destruction caused by coyotes, various treatises were published on how to trap or otherwise kill the animals. Although primarily concerned with the extermination of coyotes, these documents also provided some natural history of the species, for early trappers recognized the need to understand the habits of their prey in order to locate it. Unfortunately, the biases of the authors tainted the reliability of the information, as fact and fiction were sometimes mingled.

Today, coyote researchers continue to focus on controlling the animals, and in some western states one or more studies of predator damage are conducted annually. Meanwhile, millions of dollars have been spent developing and utilizing techniques of coyote control. In contrast, there have only been a few studies of coyotes in natural environments. In 1937, after the cessation of predator control in Yellowstone Park, Adolph Murie (1940) recorded in detail his observations of park coyotes; and in recent years, Franz Camenzind (personal communication) and Hope Ryden (1975) have engaged in independent studies of Wyoming coyotes. Beyond the boundaries of protected areas, field observations of coyotes are impeded by the difficulties most researchers experience in locating populations that are relatively free from human interference.

Probably the most versatile of today's canids, the coyote ranges throughout most of North America and in portions of Central America. Able to survive and even prosper in human-altered environments, coyote populations have expanded even as other less adaptable predators have been vanquished. From early accounts it appears that prior to European occupation of North and Central America, coyotes ranged west of the Mississippi River, from southern Canada to mid-Mexico. As early settlers became established in the Americas, they altered the environment. Trees were felled, crops were planted, and alert wild herbivores, which had long coexisted with predators, were supplanted by domestic species, animals often lacking means of self-protection. Meanwhile, populations of large predators were systematically depleted and eventually eliminated by human hunters. Much to the dismay of the pioneers, coyotes thrived in the midst of these changes. They began to proliferate in regions previously dominated by their large cousin, the wolf. Coyotes found sheep, chickens, and other small domestic animals to be an ideal food source: easy to capture, palatable, and plentiful.

Today the range of coyotes extends from Alaska to Costa Rica and from the Pacific coast to southeastern Canada and the northeastern United States. Some eastern and southeastern states probably have no resident populations, and in the northeastern states it is possible that coyote-like animals are actually coyote–wolf hybrids. Statements in this chapter refer to western coyotes; the New England wild canid possesses unique characteristics.

Throughout their range coyotes tend to avoid wet, tropical habitats, but otherwise they occupy nearly every available locality. On the mountains of Alaska, in the deserts of the southwestern United States, and even in the backyards of suburbia, coyotes make their homes.

Considered by many to resemble small wolves, coyotes are in fact physically and behaviorally distinct. Most weigh from 20 to 40 pounds;

males are on the average only slightly larger than females. During fall and early winter, when its coat is thick and long, a coyote may appear much heavier and larger than it actually is. Coyote fur is a mixture of colors, varying according to location and season. Black, brown, tan, and red tones predominate. Coyotes are further distinguished by their long, pointed noses, large ears, rangy bodies, and bushy tails.

Today's coyote is frequently seen as a solitary animal (Figure 1). Although a single family is the most typical social unit, a family often remains intact for only a few months. Courting and mating occur from December through March. From then until the birth of the pups, the mated pair will be seen together or close to one another. While the pups are young, the family is a cohesive group. But as fall approaches, family members often disperse. However, evidence suggests that although the family may disband the animals may continue to rendezvous, and the parents may even bring food to their youngsters. Sometimes two or more pups remain together after separating from the parents. Long after disbanding, the family members apparently remain somewhat friendly. When

Figure 1. Solitary coyote. (Drawing by John Slater.)

they meet, greeting may range from bare acknowledgment to wagging friendliness; in any case the aggression typically apparent when strange coyotes meet is not observed.

Multiple causes exist for the separation of families. Frequently harassment due to trapping and hunting activities is responsible. In some locations the first day of hunting seems to coincide with the disbanding of coyote families. Natural causes can also force families to split. Lack of resources, primarily food, can encourage separation and migration.

If separation occurs, it is not uncommon for the male and female parents to reunite the following year. Monogamy appears to be the favored male–female relationship. In rare cases, two whelping females and one male have been observed sharing a den. This generally has been interpreted as polygamy; but since such occurrences are unusual, it seems equally probable that a second male was the mate of the second female and he was simply not sighted or he died prior to the observation.

Pair bonds are strong in coyotes and since lasting relationships may form while littermates are young, siblings sometimes become mates even though the family disbands. If a family does not separate, then the previous year's littermates, as adults, may reside with their parents and younger brothers and sisters. If the younger animals also mate, the two litters may be raised in proximity, or even in the same den. Unlike wolves, both sexes of coyotes mature at the same pace, and even when less than a year old coyotes are capable of breeding. Factors affecting their ability to breed include actual age, diet, and general physical condition. I believe that "older" first-year adults are the more successful breeders. Actual dates of whelping cover a span of several months. Certain latitudes or altitudes may correlate with restricted whelping periods but in some areas, sections of California and Oregon for example, neighboring pups may differ by 1 to 2 months in age. In these cases, older pups seem to breed more readily as first-year adults.

After coyotes mate, the female begins her search for dens. If possible, she returns to the site of the previous year's den, often utilizing some of the same structures. Taking advantage of former coyote dens, badger holes, wolf dens, and other denlike structures, she tends to renovate existing sites rather than dig entirely new dens. The den site varies; it may be on the side of a hill or stream bank, under a boulder, or even under a building. As many as a dozen dens are cleaned in preparation for the pups. Each den may have more than one exit, and the dens may be close to one another. A water supply is generally nearby.

After a gestation period of approximately 63 days, the pups are born in one of the prepared dens. An average coyote litter consists of five to six offspring, but larger litters are not unusual. Some studies indicate that litter size may be inversely proportional to the density of the coyote population. It cannot be said with certainty which factor or factors control the size of the litter. It seems safe to state, however, that inadequate maternal diets or other debilitating conditions can prevent breeding or result in small litters. Conversely, however, even the most superb diet may not increase litter size above average.

Following the birth of the pups, the male parent's role becomes more active. Initially, he provides more for the mother than for the pups, bringing food to her while she remains with the infants. Within 2 to 3 weeks, with the eruption of their teeth, the pups are able to eat solid food. As the pups' appetites increase, both parents are kept busy supplying food for them, as well as for themselves. Small prey items may be presented to the youngsters intact, but often food is regurgitated by the parents. If the coyotes feel threatened during this period, the female, or more rarely the male, will move the pups to another den. Infrequently, the offspring will be housed in two separate dens and the parents will split their efforts between each den. This phenomenon has not been studied, but it would appear to provide added security for the pups.

When the pups are capable of controlling their locomotion to some degree, they begin to venture to the den's entrance. This usually occurs when they are 2 to 3 weeks old. Later, their parents may take them a short distance from the den to a particular area that is used for playing and sunning, or the pups may romp near the mouth of the den. If danger threatens, the parents call and the youngsters return to the safety of the den. Coyote parents are extremely protective of their young. If a pup cries in fear or pain, a parent responds immediately. When the cries are associated with intralitter encounters, response is usually limited to a quick look at the youngsters, but in cases of interspecific aggression parents will rush to a pup's aid. Human interference provides the general exception to this rule; a coyote typically avoids people even when a pup is in danger. Anecdotal situations suggest that, if possible, a parent will aid its pup after the person departs.

Generally, the female assumes a greater responsibility than the male for the pups' welfare. But after the pups begin to consume solid food, either parent will continue to raise the offspring if the other dies or is incapacitated.

When about 4 weeks old the youngsters are fairly well coordinated and extremely active, playing and fighting with one another and playing with available small prey items and inanimate objects such as rocks, leaves, or sticks. They begin to move farther from the den, and if large prey are killed nearby they will feed directly on the carcasses.

Soon the pups no longer need a den for a sanctuary. After abandoning the den the family moves about freely, using gullies, tall vegetation, and other natural sources for cover. By this time the first hierarchical arrangement among the pups has been established. No subtle cues are used by the youngsters; each pup's status is determined by its ability to outfight the others. These fights can be vehement, and wounds are sometimes inflicted. A female pup generally assumes the top position in the social hierarchy at first. Throughout the next several months, sometimes for a year or longer if the family stays together, the hierarchical positions will shift. Several alterations of order may occur, but eventually most males become physically dominant over females.

In a mated pair the physical dominance of the male may not denote the full relationship between the animals. The female is considerably more demonstrative than the male, indicating her affection with overt displays—tail wagging, licking, whining, and so on. The male is usually rather stoical as he accepts these tributes. There is no doubt in such cases that the female is submissive to the male, yet in other situations the female appears to assume the leadership role. As an example, in one pair that I have observed, the male is obviously the dominant animal. Nevertheless, it is the female who generally selects the pair's direction of travel and precedes the male along the route. Furthermore, although the female shows no concern when the male is not in view, the male becomes agitated if he loses sight of the female and does not relax until he finds her. These and other situations indicate that the role of female coyotes, although not fully understood, is complex and certainly goes beyond the common "submissive" characterization.

Whether alone or paired or a part of a family unit, a coyote tends to remain within a particular area, which is known as its home range. A coyote may occupy different portions of its home range on a daily or seasonal basis, and its home range may change during its lifetime. If persecuted, unable to find sufficient food or shelter, or perhaps possessed by wanderlust, a coyote may travel a great distance from its birthplace. By capturing, marking, releasing, and recapturing animals, data have been obtained that indicate that coyotes may travel more than 100 miles. In *The Clever Coyote,* Young (1951) summarized dispersal data gathered by the Bureau of Biological Survey in a Wyoming study. Some coyotes, recaptured 4 to 18 months after being tagged as pups, were no farther than 1 mile from their original capture location. At the other end of the scale, some animals covered 90 to 100 miles during 8 to 15 month periods. All measurements were in air miles, so actual distances traveled were undoubtedly greater.

Part of a coyote's home range may be actively defended—that is, may be considered a territory in the strict definition of the term. Certainly coyotes mark with urine their established trails and ranges. Franz Camenzind (personal communication), after observing territorial behavior of Wyoming coyotes, concluded that coyotes can be classified according to their territorial urge. Nomads are lone animals with no territories. Large aggregations of coyotes may also be nonterritorial as well as nonhierarchical, but these groupings tend to be ephemeral, existing primarily in the immediate vicinity of a large food source. Pairs of coyotes in conjunction with their nonadult young occupy territories; expanded family groups, or packs, composed of three or more adults plus various young, are also territorial. The members of the pack all observe the same boundary.

The Wyoming coyotes observed by Camenzind belong to a "natural" population; human harassment is minimal. As expected, in less placid environments territorial systems may break down. By killing and otherwise pressuring coyotes, trappers and hunters accelerate migratory movement.

Coyotes have a vast repertoire of communication signals, but when strangers meet, particularly strangers of the same sex, physical combat frequently follows. Although injuries may be inflicted, and submissive, defeatist gestures briefly ignored, the loser is, except in rare instances, able to flee. Fights between strangers of opposite sexes are considerably less violent, and the animals may establish a pair bond—assuming that neither animal is already attached to a mate. Aggression between strange coyotes is not necessarily territorial. In other words, the coyote may not be defending a specific geographical area. Basically unsocial outside its family unit, the coyote may be defending against encroachment of its "personal space," the area immediately surrounding its body, without respect to the particular physical environment. Even if this is the case, the coyote on its home ground will be more confident and will have a better chance of winning. In locations where stable territories exist, the type of fighting just described probably occurs with much less frequency. During encounters at established territorial boundaries, the animals are more likely to display, using postures and vocalizations to settle disputes.

Within the family group, an intolerance for closeness is also evident. Unless engaged in an activity that requires body contact, like greeting or playing, each coyote occupies a distinctly separate space. As an example, mature coyotes rarely sleep with their bodies touching.

Within any social group, communication without violence is highly advantageous. Mature coyotes communicate by means of vocal, facial, and postural expressions, and actual battles within social units are rare. Wounds are more likely to be inflicted during bouts of play. Yet coyotes can be possessive, and if unwilling to share its prey a coyote may bite at its mate or a member of its group. This biting gesture, however, tends to indicate threatening rather than attack-oriented behavior.

Coyote vocalizations are varied. Short, sharp, high-pitched sounds, like yips and yelps, indicate fear, anger, or pain. Growls may accompany angry or threatening behavior. The softer woof seems to denote concern or alarm. Upon greeting a mate or friend a coyote may whine or utter chirping sounds. Other vocalizations may have very specific meanings. Parents appear to command their pups vocally, eliciting particular responses with definitive calls: to come, hide, call, or be silent.

I consider howling to be the most complex coyote vocalization. Even after listening to dozens of howling sessions I am uncertain of the message conveyed by the coyotes. During most of the sessions I heard, coyotes initially responded to outside stimuli—sirens, barking dogs, and the like. On two or three occasions, pervasive excitability preceded howling. Milling, prancing, nuzzling, and pawing, the coyotes began to whine. Whining was followed by yapping, which crescendoed into that combination of yapping and howling that so completely identifies coyotes. Although they are agitated, coyotes appear to experience primarily pleasurable feelings when yap-howling. In one instance, a male coyote emitting a low, prolonged howl appeared to be distressed, perhaps due to the absence of his mate. Upon hearing his call, the female turned and looked in his direction (she could not see him), then continued unconcernedly on her way. Meanwhile, the other coyotes joined in and a robust yap-howling session ensued. Judging from the obvious sociability that coincides with yap-howling, it appears that yap-howling intensifies group cohesion.

Body posture and facial expression are also complex communication forms. A coyote uses its entire body to relate its state of being. To express joy, the ears are folded backward and the animal crouches low, body wiggling and tail wagging almost from ear to ear; its mouth is barely open with the lips drawn slightly back and the tongue is extended and retracted in a licking motion. Usually whining or chirping sounds accompany this body language. A less ebullient version of this posture expresses submission; the body is lower and the coyote in general appears apprehensive. The posture for an aggressive attitude is almost the antithesis of the submissive posture; the body is held upright and stiff, tail extended, and hair erected on shoulders and back. If expressing intent to bite, especially when escape is blocked, a coyote executes a "threat gape," fully opening its mouth with its lips drawn back, displaying its potential weapons. Typically, no growl or other vocalization is associated with a maximum threat gape.

Coyotes have a range of expression that allows them to communicate clearly with one another. Yet in many instances coyotes lead solitary existences for part of each year and have little occasion to communicate. Historically, coyotes may have assembled in larger numbers, and their communication forms may have evolved for that earlier life-style.

As mentioned earlier, disbanding may be due to direct human influences such as trapping and hunting. In addition, marginal habitats with inadequate resources encourage wider spacing of coyotes. Yet, alternative causes for disbanding exist. Parent animals that were forced to disperse when young may in turn scatter their pups at an early age. Another possible cause of dispersal involves artificial selection. Certain coyotes naturally prefer socializing; others tend to be solitary. This represents a normal range of behavior. Socially oriented coyotes are generally more conspicuous and thus more susceptible to human depredation. Killing of these coyotes would result in a population possessing a preference for less socialization.

If the family unit dissolves, then each member must be autonomous. Young coyotes, although not yet adult in size, are already self-sufficient. An opportunist, a coyote will eat fruit, insects, birds, or mammals—with an occasional bite of something more unusual. Initially, the young coyote hunts insects and small rodents. An innate behavioral pattern is followed: the coyote leaps into the air, then lands with its forepaws trapping the prey against the ground. The coordination for this move is perfected during games and early prey captures while the youngster is still under the protection of its parents. Skills involved in capturing larger prey, like sheep and deer, may require more learning. Confronting an animal larger than itself, an inexperienced coyote is more likely to flee than attack. When an inexperienced animal does attack, it may direct random, sometimes ineffectual bites at the prey. Experienced coyotes, on the other hand, utilize a moderately standard throat grip, biting the underside of the throat and causing death by suffocation. Some learning, whether by observation of another coyote or by trial and error, seems to be necessary in the killing of large prey.

Detailed studies of coyotes' food preferences have been conducted in diverse settings. Unfortunately, each survey is potentially obscured by its innate biases, including (a) study area location, (b) date of collection, (c) funding source, (d) types of analyses (fecal or stomach contents; direct observation), and (e) professional status of collectors (trappers, livestock owners, researchers, and so on). Rather than discuss each study, I will briefly summarize the findings.

Essentially, a coyote eats whatever prey is easiest to obtain, with selection if possible for more palatable foods. Food availability is determined not only by the numerical status of the prey species but also by the coyote's ability to capture that particular prey.

Under temperate conditions coyotes mainly capture rodents and rabbits. In snow-covered areas, coyotes may dig for rodents; if the snow is deep, yet crusted, large herbivores that flounder and die provide carrion for coyotes and other scavengers. Snow can also be the coyote's nemesis. If snow is shallow enough to allow easy passage of herbivores, yet deep enough to impede the coyote's progress, the coyote may die from starvation. Of course livestock provides another food source; individual coyotes may prey heavily on unprotected sheep, chickens, and turkeys.

Insects are sometimes a large part of a coyote's diet. In fact coyotes seem to relish certain insects, as well as other small invertebrates. I have watched a young coyote hold a snail between its front paws, then shell and consume the snail with as much pleasure as a connoisseur. Grasshoppers also appear to be a favorite food and when present in large numbers they are readily consumed. Coyotes catch grasshoppers in flight or trap them against the ground with their forepaws. Coyotes spend hours in pursuit of grasshoppers, sometimes pausing between catches to chew in an exaggerated, prolonged manner before swallowing.

Although meat is the major component of the coyote's diet, fruit and grass are also important foods. Wild and domestic fruits are usually eaten by choice, not merely as a last resort to avoid starvation. Examining one healthy young coyote, who had been killed because he erred in selecting a path that traversed a sheep pasture, I discovered only manzanita berries in his stomach. Other fruit-eating coyotes have been caught raiding watermelon or other fruit fields.

This list of foods is by no means all-inclusive. A complete compilation would include a variety of other mammals, birds, fish, reptiles, invertebrates, and vegetation that are less frequently eaten. According to Murie's (1940) findings, it would also include rags, gloves, curtain material, cellophane, tinfoil, rope, and much more!

Currently coyotes have little need for natural regulatory mechanisms over most of their range. Human depredation overculls coyote populations. But under natural conditions coyote numbers are limited by starvation, disease, and parasitism, as well as by inter- and intraspecific strife. Occasionally the predator becomes the victim; deer, elk, and antelope are capable of attacking and killing coyotes. Intraspecific interactions may indirectly affect coyote populations by impeding breeding, thereby regulating population size by altering the birth rate. Social interactions are particularly significant when coyotes reside in groups. Interference by an older coyote, for instance, may prevent the mating of a younger animal. On the other hand, an immediate decrease in population size results when coyotes kill other coyotes, as occurs, for example, when adults kill unrelated pups.

Quick and agile, an adept hunter or hidden observer, a coyote is well adapted for survival (Figure 2). Its supple, muscular body is further aided by acute senses of vision, hearing, and smell. This physical prowess is matched by an alert intelligence. Coyotes appear to enjoy puzzling situations. Possessing seemingly unlimited patience and perseverance, they willingly seek solutions to problems they encounter.

By utilizing these faculties, lone coyotes are able to devise systems enabling them to maneuver their prey into vulnerable positions. The following account of a captive coyote illustrates this ability. Although the coyote was fed a commercial diet, chicken feathers were found in the coyote's pen on several occasions. Chickens wandered freely on the surrounding grounds, but it was difficult to believe that a chicken would willingly enter an enclosure containing a coyote. To discover the solution to the mystery, an observer watched the coyote from a hidden location.

Figure 2. This young female coyote, confined in a large fenced enclosure, appears tense, but not cowed, when her human captors are near. (Photograph by Roberta L. Hall.)

After the coyote received his usual meal, he placed several large chunks of his food near the fence of his pen. The coyote then hid. Soon several chickens began to peck at the food. Eventually, a chicken selected the morsel closest to the coyote. Patiently waiting until the chicken was engrossed in its meal, the coyote pounced, seizing the chicken by its neck. Still ignoring the food remaining in his dish, the coyote began to eat his preferred dinner!

Numerous eyewitness accounts of cooperative hunting by coyotes have been recorded. Some appear incredible—yet, considering the physical and mental capabilities of coyotes, these accounts should not be dismissed lightly. Coyotes use similar techniques when hunting antelope and jackrabbits, despite the disparity in prey size. One coyote initiates the chase while the other positions itself along the chase route and waits. As the prey dashes by, the predators switch roles. Alternately resting and chasing, the coyotes work as a relay team. Obviously this method works only if the coyotes are able to anticipate the behavior of the prey. Exhausted and terrified, both jackrabbits and antelope tend to circle, and the coyotes capitalize on the fear-induced reactions of their prey.

Coyotes also exhibit teamwork when one member of a pair flushes an animal and the other captures the fleeing quarry. Many observations of this technique have been made. For example, one coyote was seen scratching at a wood rat's nest and, as the rat attempted to escape, the second coyote, which had been silently observing, caught the rat. An additional example of joint effort is seen when two or more coyotes cooperate by simultaneously attacking an animal; this technique may be successful with large prey species.

Another means of teamwork involves distracting the prey. One coyote blatantly displays itself, often cavorting about—bounding, dashing, and rolling on its back. The other remains hidden until the fascinated prey is drawn into range; then the hidden coyote strikes. A single coyote may attract prey by the same method. If the prey moves within range, the coyote abruptly ceases its capering and pounces. Most accounts of the fascination display describe coyotes enticing birds, but undoubtedly other animals are also beguiled by this maneuver.

In a manner reminiscent of man's association with dogs, coyotes sometimes form relationships with members of another species. Although the ramifications of the relationship are not well understood, coyote and badger pairs have been sighted often enough for their occurrence to be regarded as more than freak happenings (Figure 3). Both animals eat small rodents and both dig after burrowing prey. Perhaps a coyote and a badger cooperate in the same manner as a pair of coyotes, and as one animal flushes the prey, the other captures it.

Figure 3. Coyote and badger hunting together. (Drawing by John Slater.)

Unless protecting pups, coyotes adhere to the familiar adage that discretion is the better part of valor. If cornered, a coyote will fight desperately and ably, but when given a choice coyotes react to dangerous or frightening situations by fleeing. By relying on their speed, small stature, and quiet maneuverability, coyotes are often able to avoid detection.

Despite their limited ability to manipulate their environment, coyotes have evolved into an exceptionally versatile and prosperous species. In fact, if one wished to create an ideal predator, the coyote would probably be selected to serve as the prototype. But, even with their extraordinary physical and mental capabilities, coyotes possess no immunity from human depredation, and the future of this small predator is by no means secure.

REFERENCES

Murie, A. *Ecology of the coyote in the Yellowstone*. National Park Service, Fauna Series 4. Washington, D.C.: U.S. Government Printing Office, 1940.
Ryden, H. *God's dog*. New York: Coward, McCann, and Geoghegan, 1975.
Young, S. P. *The clever coyote*. Washington, D.C.: Wildlife Management Institute, 1951.

4

Comparative Ethnology of
the Wolf and the Chipewyan

Henry S. Sharp

Throughout our cultural history we members of Western civilization have been subject to a society that has strongly separated human from animal. This distinction is so ingrained in our culture that even scientists and philosophers find it difficult, if not impossible, to avoid unconscious assumptions about the nature of the very subjects they investigate. Attempts to deal with these cultural statements, which are values, have largely been failures. Even as brilliant a theorist as J. B. Watson, drawing upon the path-breaking attempts of Thorndike, fell victim to these assumptions in the formulation of behaviorism. He pushed his analysis to the opposite extreme in developing a scientific position that denies the similarities between human and animal and avoids the consideration of animal mental processes whenever possible.

One fortunate by-product of this cultural heritage has been the emphasis upon culture as a mediating process between man the spiritual being and man the animal. It is this area, the analysis of culture as a mediator, that is the home of social anthropologists. As we attempt to extend their analyses to the behavior of social animals we find the concepts and techniques of a discipline that has ignored the behavior of animals until recently. Many of the tools of social anthropology will prove inapplicable to animal social behavior, and more will require extensive modification before they can be used, but social anthropology does provide a starting point directed to the analysis of social rather than individual behavior.

This aspect is of greatest significance, the social anthropologist regards culture as a *social fact*, a process separable from the population that bears it and independent of the conscious plans of the species that bears the culture. Culture is given primacy as a causal factor in the regulation of behavior. Culture determines the utilization of the environment to meet the biological needs of the species. This is a point too easily forgotten, for our species, *Homo sapiens*, is an animal: a mammal, nothing less and nothing more.

Mammals are bound by certain biological requirements that must be met if the population is to survive, but these requirements, which I shall label the *functional requirements* of culture, are rather minimal. It is easy to interject values such as self-fulfillment, comfort, and quality of life into an analysis of cultural systems, but it is only necessary to ensure food, shelter, and reproduction for some percentage of the culture-bearing population in order to maintain that population at a suitable level, thus providing for the continued existence of culture. Beyond the functional requirements of culture all other aspects of social behavior must be viewed in terms of the "logic" or "rules" of each cultural system.

The implications of this theoretical position are significant but few have been systematically investigated and I do not presume to give any final answers here. This chapter is a limited, nevertheless comparative, ethnological analysis of the Chipewyan *(Homo sapiens)* and the arctic wolf *(Canis lupus)*. I shall explore some of the implications of the subsistence choices that are common to both.

These two populations are hunters of barren ground caribou and their social systems are adjusted to the hunting of barren ground caribou. Because caribou are absent from the ranges of the Chipewyan and the wolf for part of the year, the two groups of hunters also depend on animals other than caribou. However, the cultural decision to hunt caribou as the primary item of subsistence has produced remarkable similarities between the two species in their utilization of land and in the formation and distribution of social groups. The cultural decision to hunt caribou results in a population density lower than what would result through other decisions regarding the utilization of resources, since the emphasis on caribou hunting often leads to the "inefficient" use of resources by both Chipewyan and wolf.

The basic choices regarding subsistence patterns, social organization, demography, terrain usage, and yearly cycles are made on the basis of the internal logic and structural characteristics of the two cultures under analysis. Therefore ecological, psychological, behaviorist, or other analyses that exclude sociological and cultural phenomena are necessarily incomplete and cannot uncover the primary factors determining the behavior of social animals.

THE REGION OF ANALYSIS

This analysis is limited to the area utilized by the Stony Rapids band of Chipewyan. This limitation reflects the limits of my fieldwork[1] and the existence of more data on the Chipewyan than on the wolves of the area. The region in question is a block of boreal forest, transitional boreal forest, and tundra in Saskatchewan and the Northwest Territories covering over 40,000 square miles. Within Saskatchewan the area is bounded by a vertical line bisecting the province midway between the towns of Stony Rapids and Fond du Lac and continuing south to the north end of Cree Lake. The eastern boundary is a vertical line equidistant between the settlements of Black Lake and Wollaston Lake. North of these towns the boundary turns northeast toward the Saskatchewan–Manitoba–Northwest Territories border and continues into the Northwest Territories to include Snowbird Lake, Kasba Lake, and the northwest section of Ennadai Lake.

The approximate southern boundary connects the two vertical lines just north of Cree Lake. The western boundary continues from the Northwest Territories boundary approximately 20 miles west of Selwyn and Smalltree lakes to the western end of Knowles Lake. At present the Chipewyan population ventures no farther north than the line between the south shore of Damant Lake and the northwest corner of Ennadai Lake, though currently active trappers have worked as far north as Gravel Hill Lake and Mosquito Lake. In the nineteenth century the Chipewyan penetrated the barren grounds at least as far as the calving grounds of the Kaminuriak and Beverly caribou herds (see Figure 1).

The three zones shown in the figure correspond to the boreal forest (1), the transition zone of the boreal forest (2), and the tundra (3). At present the Chipewyan exploit zones 1 and 2 in a pattern of movements with maximum dispersion during the fall and maximum concentration during the summer. This pattern reflects the post-World War I emphasis upon trapping and particularly the post-World War II decline of the caribou herds and the concomitant collapse of the fur market.

[1] The fieldwork upon which this chapter is based was conducted in 1969–1970, 1972–1973, and 1975. A total of approximately 2½ years was spent in the field, of which approximately 11 months was spent in the bush in all ecological zones. The longest continuous period in the bush was in 1975 at tree line and lasted for 8 months.

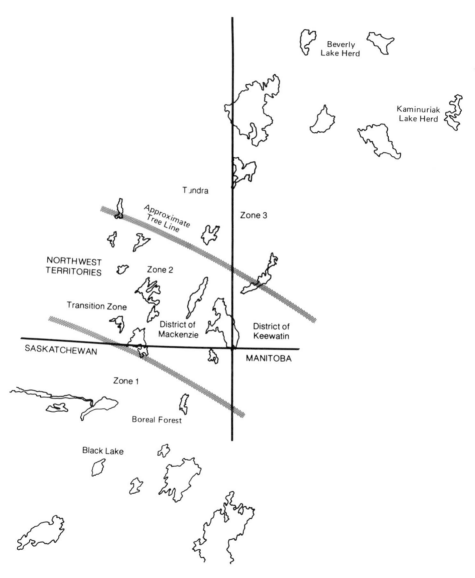

Figure 1. Stony Rapids Chipewyan traditional area. Zone 1 is boreal forest; zone 2 is transitional boreal forest; zone 3 is tundra.

THE CHIPEWYAN AND THE CARIBOU

Through the use of older Chipewyan informants it has been possible to reconstruct the nineteenth-century pattern of movements that was less oriented toward trapping and the fur trade. It is this pattern that I shall describe in detail, as it was directed primarily to the pursuit of the barren ground caribou.

Prior to discussing the movements of the Indians and the wolves, I should briefly discuss the movements of the barren ground caribou. Two herds, the Beverly and the Kaminuriak, enter the region under discussion. Both herds are found on the high barren grounds in June during the calving season. The caribou wander considerable distances during the summer but they regularly appear at the edge of the trees between the end of July and mid-August. After approximately a month of circling movements they penetrate the transition zone (between mid-September and early October) as the small lakes freeze and take on snow cover. Caribou movements are extremely unpredictable but ideally the animals continue south with freeze-up and reach Selwyn Lake by October and the Lake Athabasca–Black Lake area by mid-November. In the past the bulk of the herd may have continued south of Lake Athabasca for some distance (Bone, Shannon, & Raby, 1973), but a significant portion of them wintered in the region between Black Lake and Selwyn Lake, as is the case at the current time.

The caribou remain in or near this area until the weather breaks in early spring, when they begin a rapid migration to the calving grounds. The movement of the pregnant cows is more purposeful than that of the bulls, as bulls are not as attracted to the calving grounds. This migration, which could begin as early as March, brings a stream of caribou through the Chipewyan hunting units from early spring until just before breakup of the lake ice.

The geographical movements of the Chipewyan were directed toward the taking of large numbers of caribou during the migrations, as it is during the migrations that the caribou are most vulnerable; they travel in bunches and are not greatly concerned by predators. They are easy to approach and are to a certain extent canalized in their movements by weak ice, narrow areas of open water, and rough terrain. The general movement of the Chipewyan was from a dispersed pattern of summer subsistence activities in zones 1, 2, and 3 to a concentration of population in zone 2, in order to exploit the caribou's vulnerability during the migrations. Depending upon the success of the caribou kill during the fall migrations through zone 2 (and the desire to trap during the late contact period [Helm & Damas, 1963]), part of the Chipewyan population moved south into zone 1 to winter with the caribou. This pattern of movement was inverted with the spring migration, the Chipewyan moving north and dispersing for the summer.

As the caribou migrated north in the spring, the Chipewyan population that chose to remain in zone 1 for the summer moved to creeks and river mouths so they could fish during breakup, and then dispersed into the bush for the summer.[2] The Indian population of zone 2 took caribou at strategic points for dry meat but a significant portion of the population moved out onto the barren grounds to locations where they had access to the caribou. The Chipewyan of zone 2, though more dependent upon caribou dry meat than the Chipewyan of zones 1 and 3, also appear to have dispersed—however, the evidence from informants and surface inspection of old settlement sites is less than conclusive. The population that was permanent in zone 1 depended upon the moose and woodland caribou but was concentrated in the Churchill River damage.

The post European contact period pattern of movement described for the Chipewyan began to break between the mid-eighteenth century and the end of World War I. During this period the yearly cycle of the Indians began to adjust to a lesser dependence upon the caribou and greater concern with Western trade goods, foods, and wage labor. By the late 1930s the pattern had adjusted to a basic north–south movement; zone 3 was no longer used and the Chipewyan concentrated on the taking of caribou during the migrations. There was relatively little movement of entire families during the winter and many families summered in zone 1. It was not until the post-World War II period that a significant number of trappers spent a large part of the year around the Hudson Bay Company stores and not until 1974–1975 that trapping and wintering in the bush ceased to be a significant activity (Sharp, 1975a,b).

[2] By the mid-1920s a small but consistent number of Chipewyan were spending the summer and part of the rest of the year near the trading post of Stony Rapids. As early as the mid-nineteenth century some Chipewyan spent part of the year around the post at Fond du Lac and in the eighteenth century the Chipewyan expanded into the boreal forest as far south as the Churchill River drainage. Part of this population was drawn to the stores along the Lake Athabasca–Black Lake line during the nineteenth century but their pattern of movement was not based on the caribou and was not part of the aboriginal pattern. It is very difficult to uncover reliable statements about their movements during the period of analysis (contact traditional period) but the boreal forest Chipewyan seem to have been few in number until the 1930s. See Sharp (1973, 1975a,b) for a discussion of the internal structure of the Chipewyan.

Chipewyan Social Organization and Caribou Hunting

The adoption of the position of cultural determinism, indicated at the beginning of this chapter, necessitates the view that social organization, food distribution, and subsistence practices of the Chipewyan are intimately related structures. The interrelationship of these structures is such that no single one of them can be regarded as basic. Changes in any of the subsystems must be reflected in the other subsystems and those changes, as a response to the initial change, tend to be homeostatic; that is, they tend to preserve the nature of the original system. I wish to stress that the nature of these interconnections is not analogous to a simple mechanical linkage in which a deflection in one system produces a direct deflection in the others. The interconnection should be viewed as one of dynamic balance. Changes in one part of the total system are ultimately met by changes in other parts that tend to minimize total change. Considerable room for variation and response exists within the entire system, which allows for manipulation and variation, inherent parts of any cultural system; but the changes are modulated by the cultural system.

The Chipewyan subsistence and food-distribution systems, as well as their social organization, find expression through the kinship system. Kinship is the primary vehicle by which people are linked into the social groups that provide and distribute food. As my primary concern is a comparative ethnology between the Chipewyan and the wolf I shall stress the distribution of social groups within the social system rather than the working of the kinship system that provides a mechanism for the distribution.

The social organization of the Black Lake Chipewyan at the time of contact with the white man is problematical and controversial. The controversies need not concern us here, as all views predicate the existence of localized groups that cooperated in subsistence matters through the use of pounds, drives, and equivalent fishing techniques such as weirs. In this chapter I shall assume that the contemporary social organization is nearly the same as the social organization at the time of contact, since the cultural strategies involved in hunting caribou and the basic ecological situations were quite constant until the influx of white trappers after World War I.

The Hunting Unit

The Chipewyan hunting unit is a bilateral kinship grouping that consists of a nuclear family plus unattached children, aged dependents, and as many nuclear families attached through descent or marriage as can be recruited. At the present, with trapping a factor because of its economic significance, the upper limit in size appears to be 17 to 20 persons, of which up to 4 can be adult males of various ages. Structural features of the kinship system make this number appear to be a reasonable upper limit for the earlier period. A hunting unit may occupy a cluster of several separate dwellings (tents or cabins) but is always a social group within which food, equipment, and labor are shared without cost. (''Cost'' here is not an economic statement but a statement of the loss of prestige at having to ask for food, for example, and is an explicit statement of the failure of the supplicant's magical power, which in turn produces a loss of status.) All residential groupings in Chipewyan society consist of multiples of hunting units. These groupings may consist of a single pair of hunting units sharing a base camp (but keeping their economic and subsistence activities separate) during the winter, up to the ''villages'' of several hundred people, though the duration of these ''villages'' during the contact period and earlier is uncertain.

The ability to form interlocked residential groupings from discrete corporate groups without sacrificing the corporateness of the groups has obvious sociological implications and underlies all aspects of Chipewyan society. This feature provides the basic mechanism that allows the Chipewyan to exploit an unreliable animal species. This aspect of Chipewyan life is apparent to even a casual observer provided he realizes that Chipewyan society functions as a system that includes hunting units as elements, just as a hunting unit includes individuals as elements.

Chipewyan Strategies of Caribou Hunting

Like systems of philosophy and religion, cultural systems focus on simple problems that are given profound implications. Chipewyan culture is no exception. Given the cultural decision to be caribou hunters above all else, then it is incumbent upon the culture—if it is to survive—to provide enough caribou to meet the functional prerequisites of the Chipewyan Indians as biological organisms. This is difficult, as caribou are not predictable in their movements at a local level. They do not always come at the same place or at the same time and the problem is compounded by the fact that the land area the Stony Rapids Chipewyan occupy is immense and the human population is low. (In 1975, even with a probable doubling of the Chipewyan population in the last two generations, the population density is only one person per 65 square miles.)

Within any small geographic area, regardless of how the boundaries are derived, there is no certainty as to when and where the caribou will appear. The smaller the unit of area the more critical this factor becomes.

Given any area, as the size of the area is increased there comes some point at which the probability of arrival of the caribou within that area is sufficiently large for a social system to operate with the certainty of the caribou appearing. In other words, the Chipewyan must have an adequate number of hunters wherever the caribou arrive at whatever time they arrive to ensure an adequate kill of caribou. This enables the minimum level of population within the bounded region to survive through subsequent sharing of the caribou.

The Chipewyan have solved the problem through a pattern of simultaneous concentration and dispersal of hunting units. Hunting units are concentrated where there is a high probability of the appearance of caribou. Hunting units are dispersed, singly and in small multiples, where caribou are less likely to appear either at all or in large numbers. This pattern of simultaneous concentration and dispersal is not consciously planned. It is derived from the internal working of the Chipewyan social system, which constrains the behavioral options and the perception of behavioral options of the Chipewyan. Hunting units space themselves for reasons of ambition, preference, ignorance, conflict, curiosity, and sentiment, among a host of other reasons generated and conditioned by the cultural system. They do not place themselves on the basis of conscious perception of the operation of the cultural system or coordinated planning. The means by which a cultural system ensures its survival *cannot* be left to the conscious understanding of the system by individuals within it.

The pattern of simultaneous concentration and dispersion ensures that no matter when or where the caribou appear one or more hunting units will be waiting to exploit them. A single hunting unit in isolation that finds itself in a high caribou density area will kill caribou by the hundreds (by my field observations, a single hunting unit in isolation among the caribou will kill caribou at an approximate rate of 216 caribou per person per year).[3] They can provide surplus caribou to support a large population for a long period of time. This is especially true when the caribou are processed by the Indians with the intent of stretching supplies.

This strategy has one obvious drawback. It ensures that some hunting units will always take enough caribou to keep the system functioning but also ensures that some hunting units will not take an adequate supply for

---

[3] Though the Chipewyan do not have access to the caribou for a long enough period to reach a figure of 216 per person per year, the caribou kill of a single hunting unit with a maximum population of 11 including the ethnographer was over 200 in 5 weeks from July 29 to September 11 and claimed another 100 during the fall after my departure. In 1975 a group of Indians that did not exceed 16 took over 250 between late August and early November but another 200 (minimum) were killed by transient Chipewyan hunters flying in and out for short periods; *no more* than 40 of those killed by the transient hunters were utilized for food.

their own needs. In either case some hunting units (sometimes most hunting units) must depend on the surplus kills of other hunting units in other locations. Both situations reinforce the existing pattern of hunting unit corporateness and the kin connectedness that provides the exchange system binding the cultural system together. The subsistence pattern that provides adequate caribou to keep the system functioning also reinforces the interconnectedness of the hunting units that makes the simultaneous pattern of concentration and dispersal possible.

THE WOLF

The wolf population in the study region has not had the "advantages" that have been bestowed upon the Chipewyan by the Hudson Bay Company, the Roman Catholic missions, and other interested agents of Western civilization. Although the wolves have been objects of much concern in the fur trade they have not been subjected to the influx of disease, exploitation, and moral concern to which the Indians have been subjected. In all probability the wolf population has managed to retain more of its aboriginal social structure into the twentieth century.

By the end of World War I, when large numbers of white trappers appeared in the region, the wolf population entered a period of intensive contact with Euro-Canadian civilization. The trappers, who were common until World War II, brought direct and indirect pressure upon the wolf population. The direct pressure came through the shooting and poisoning of wolves, either for their pelts or as part of a misplaced symbolic equation of the wolf with forces disruptive to human control (see Introduction). Few wolves were trapped but large numbers were shot (Figure 2). An inestimable number of wolves were destroyed by poison. Informants, both Indian and white, say that the wolf population was greatly reduced and is still low but it is doubtful that direct human predation is entirely responsible for this decline.

Wolf populations have shown a remarkable capacity to recover once direct predation (for example shooting, poisoning, digging up dens) is terminated or reduced (Rausch, 1967). Wolf populations are not quick to recover from the indirect pressures Euro-Canadian society brings to bear through environmental modification. In the study region the primary effect of the contact period upon the wolf population has been the reduction of the caribou herds, which provide the primary subsistence item for the wolf. The decrease in the number of caribou has resulted in fluctuations and reductions in wolf numbers and a possible—but unknown—shift in social organization and subsistence patterns to compensate for the re-

Figure 2. Wolf kill. This adult male arctic wolf was shot for its hide. The carcass was hidden in the bush and the mouth propped open with a stick so the hunter would remain in favor with the wolf. (Photograph by Bernadette Sharp.)

duced primary food source. Fortunately wolves are not subject to welfare and government sedentarization policies. The wolf population has probably managed to retain its primary features of social organization and subsistence activity, though with increased governmental intervention this soon may not be the case.

A complete discussion of the wolf population would require the area of analysis to be extended into the barren lands at least as far as the calving grounds of the caribou, a range no longer exploited by the Chipewyan. My analysis, since I have accepted human boundaries, will be less than complete for the wolf population; nevertheless the most salient features of the wolf social system will be discussed.

Unlike Chipewyan social groups and individuals, the wolf population does not pass through a single geographic point. This makes it impossible to identify all the individuals and social units in the region, but it also prevents us from (erroneously) regarding the wolf population as being defined by those members of the population that are seen at one particular location.

This is advantageous to us as analysts as it forces us to recognize the complementary roles played by the processes of dispersal and localization even if it does not reveal the social dynamics and/or psychological forces that result in these processes. I suspect that these two processes have as their basis the contrasting requirements for pup rearing (which requires dispersal) and for recruiting adult wolves into established social groups (which requires localization).

The forces that generate the process of dispersal for pup rearing operate at a social level to ensure adequate (though "adequate" is not measurable by any single factor such as number of pups) utilization of the environment for production of pups and new social groups. These forces include, at a sociological level, all those factors that produce colonization of new areas and recolonization of old areas as well as intrusion upon the areas occupied by less successful social groups. The factors that produce localization of the wolf population operate at a social level to allow existing social groups to recruit compatible individuals, so that the group can maintain themselves as social groups; they also allow for the formation of pairs that are the nucleus of new social groups. The process of group formation during the localization phase allows the entire system to adjust the number of internal social groups as an adaptation to local temporal conditions, and provides a reservoir of groups for colonization of new areas, recolonization of old areas, and intrusion into areas that are occupied by sociologically inferior groups (Mech, 1973). The two processes have similar sociological functions in some aspects, but the dispersal phase provides minimal opportunity for the formation of new groups and maximal production of new individuals (by birth), whereas the localization phase provides minimal production of new individuals (by recruitment) and maximal production of new social groups.

According to the postulated commitment to the causal role of culture made early in the chapter, the social systems of the wolf must meet the functional requirements of the individual wolves. The shelter and reproduction aspects of the wolf social system are apparent to a human observer since they occur at what seems a rapid pace to a longer-lived species. (The life span of the wolf is about 10 years; that of a human, about 70 years.) The difference in life span is deceptive, however, as it tends to give the reproductive process (including shelter in this rubric, as the constitutional characteristics of the wolf make physical shelter relatively unimportant except for short periods after parturition) a rigid, calendar-determined, and mechanical appearance. The wolf can produce an entire generation in about 3 years, the approximate time from birth to social maturity of a male wolf. Proportionately (3/10) this corresponds well to the time (20/70 or 25/70) of a human generation, both as a reflection of the time to social maturity from birth of a human male and as a percentage of normal maximum life span of a male of either species.

As I am concerned here primarily with the joint adaptation of Chipewyan and wolf to the exploitation of the barren ground caribou, I shall examine in greatest detail how the subsistence activities of the wolf are interrelated, since these activities are shaped by its social structure to meet its minimum functional requirements and the environmental possibilities available.

THE WOLF AND THE CARIBOU

The terms *localization* and *dispersal* were used in the previous section to avoid too obvious a loading of the argument but I hope that they can now be seen as variations of the Chipewyan "simultaneous pattern of concentration and dispersal." The manner of implementation of the strategies, and the strategies themselves, are slightly different for each species. Both are variations of the same process, however, and are used to meet the differing functional requirements of the two social systems.

For the wolf population the period of maximum dispersal occurs in the late spring and early summer during the period when the pups are born and the social group is tied to a specific geographic area containing the den or dens. The maximum period of dispersal is relatively short if we regard the move to a rendezvous site as a break away from a specific geographic area that results in increased social contact between wolf social groups (Joslin, 1966). It corresponds to that period of the year when caribou are absent from the largest geographical portion of the area of analysis, and the beginning to middle of their longest absence.

During this period the caribou are concentrated in zone 3 and are effectively absent from zones 1 and 2. For purely ecological and subsistence reasons one would expect a concentration of dens and wolf social groups around the calving areas where subsistence, especially by the taking of calves, is easiest. Yet the wolf population, which dens in all three zones (Kuyt, 1972), is at its greatest dispersal. If this dispersal, when the caribou are most vulnerable, has long-term ecological advantages for the survival and numbers of the caribou herds and indirectly for the wolf population, this cannot be apparent to the individual wolves or wolf social groups. When a social animal disperses away from its primary food source at the time it rears its young and is vulnerable because of the extra demands of the young, we can only conclude that this action is the result of forces within the social system.

The wolf social groups denning in zone 3 are better able than the populations in zones 1 and 2 to exploit the caribou directly throughout the denning period. Those in zones 1 and 2 must turn to alternative food sources. I shall discuss some of these alternative sources later but shall consider the role of the caribou in this section.

It is fairly well established now that wolves will take prey in excess of their immediate needs (Mech, 1970). This is a behavior that has been interpreted as useful in maintaining not only the wolf but also secondary predatory and scavenging populations (for example, foxes and ravens). Indian informants are aware of this aspect of the wolf's excess kills but they attribute to the wolf sufficient foresight to kill an excess of caribou near the den site in order to have an adequate food supply when the caribou are absent. The process of taking excess kills has been reported in the literature, and also by whites in the study area (personal communications of C. Terry and F. Riddle), without any implication that planning is involved.

We need not consider whether or not the wolf engages in purposeful behavior (which is quite a different question from whether or not there is a wolf culture) in the sense of planning activities directed toward conditions absent in time or space from sensory perception in order to see that the attribution of purpose by the Indians is significant. They think the killing of excess caribou during the spring migration is a mechanism to ensure an adequate food supply during the denning period. This description of the situation by the Indians is correct as a description of events and can be explained *without* reference to the motivation of the individual wolves.

At the beginning of the spring caribou migration the wolves have usually just completed the breeding season and are concentrated around the caribou herds. As the caribou begin to move north they become increasingly vulnerable to wolf predation as a result of the loss of wariness that occurs when the caribou form large bunches and begin to move. Wolves are able to kill caribou in large numbers, but the caribou move fast in the spring migrations and the wolves are not able to utilize the caribou they kill in excess of their immediate needs. Those wolves that are returning north must remain within a reasonable distance of the caribou herds during this period if for no other reason than that they must reach their denning area in time for their own pups to be born.

The wolves thus begin their period of dispersal with the migration of the caribou, but as a result of their own dispersal to their denning areas they cannot use all the caribou they kill. Thus the unused caribou become a resource for long-term utilization by the component subgroups of the wolf population—that is, by the breeding packs that remain in the area.

In the initial stages of the migration of the caribou, the bulk of the wolf population is in effect providing a reservoir of food supplies, not only for subsidiary species (fox, raven, and so on) but also for its own subgroups. As the migration moves northward progressively fewer wolves are following the caribou, as family packs that den in available territory along the route drop out. Fewer excess caribou carcasses are deposited along the way for the pup-rearing packs, but this diminution is a product of the diminishing distance to the calving grounds and is related to the shorter duration of the absence of the caribou. This is particularly noticeable in areas close to tree line where the lack of surplus kills during the migration is compensated for by the erratic availability of small groups of caribou, particularly bulls, that do not complete the journey to the calving grounds, or are moving at a relatively leisurely pace.

When viewed from this level of analysis, the migration illustrates how the actions of the entire wolf population affect the survival of constituent groups and how the apparently pointless taking of "excess" caribou is structurally conditioned to provide an essential feature—dispersion—for successful reproduction of the population. Since the migration routes taken by the caribou are variable this behavior may be a partial explanation of the nonuniform utilization of the wolf's range for denning.

The concentration phase of the wolf's yearly cycle begins at the time the wolves break away from a den site and move to a rendezvous area. The wolf's inherent capacity for mobility is brought into play progressively from this point onward until the height of the concentration phase about the beginning of the breeding season. With the increased mobility and the break away from restricted ranges come increasing social contacts—contacts that must be met with a minimum of intergroup hostility if the wolf population is to maintain itself as a system.

If the fall migration operates as an inversion of the spring migration there might be enough caribou carcasses left on the tundra to allow a population of wolves to winter there. I lack adequate information to make such a statement and unfortunately the published reports (Kuyt, 1972; Parker, 1973) on the area are biased by the excessive wolf kill of a white wolf-poisoner. My field investigations indicate his kill of wolves is in excess of 300 per year, of which no more than half were picked up during the period 1968–1973, at which point significant wolf poisoning appears to have been thwarted by governmental agencies and the advancing age of the wolf poisoner.

The end of the period of dispersal of the wolf population is keyed to two phenomena: the development of the wolf pups and the movement of the caribou as they enter the fall migration (Figure 3). The first factor allows the wolves to be mobile and the second factor requires the wolves to be mobile.

The arrival of the caribou at tree line brings a following wolf population

Figure 3. Caribou cow in early fall. (Photograph by Bernadette Sharp.)

that is already at a greater population density than during the period of dispersal. On the ground at tree line it is impossible to gather much information about the wolf movements; air support is necessary. Indirect indications of wolf activity (audible howling, tracks, sightings, behavior of sled dogs) increase shortly before the arrival of the caribou, diminish during their passage, increase for a short period after the passage of the caribou, and then drop off over a period of a week or two.

The observed data may be interpreted in several ways, but two models seem particularly plausible. The simplest solution is to regard the concentration of wolves in front of and behind the migrating caribou as consisting of the wolves that denned on the tundra following the caribou, whereas those that denned within the transition zone concentrate in front of the migrating caribou herds. The wolves concentrate several days before the arrival of the caribou so this explanation requires the analyst to assume the existence of a communication system that functions with the linguistic feature that Hockett called *displacement*. (Displacement in a language system is a reference to things not present, either in time or in space. The presence of this feature in a communication system is critical, as more than any other feature it is a clear and readily observable sign of a symbolic system.)

The second possibility is to regard the wolves as not belonging to distinct groups based on summer location. This view regards the wolves as dispersed by small groups derived from pup-rearing packs that are separately engaged in the pursuit of caribou and move over large areas ahead of, to the side of, and behind the caribou herd seeking out easy kills. This view does not require any statement about the wolf's communication system and fits with the method of utilization of caribou kills. Both these explanations preclude the possibility of wolf social behavior being based on fixed territory during this phase of the yearly cycle.

From the time of their arrival at tree line until their departure for the south, a period that can cover 3 months, the caribou move continuously but slowly in large circling movements. The wolf population can easily make contact with the caribou herds and the wolves movements appear, from the ground, to be relatively nondirected and erratic. Pack size seems consistent with those sizes reported for pup-rearing family packs elsewhere, and there is no indication of the formation of large packs of wolves at this time of the year either from my observations or from informants.

Once the caribou begin to move south, normally when the small lakes are frozen and have a snow cover of at least 3 inches, they move at a rapid pace but make large circling patterns. The majority of the wolf population follows the caribou within a week of their departure but some wolves remain at tree line at least through Christmas. They appear to subsist on the excess caribou kills of both wolves and men and on the few groups of wide-ranging caribou. A significant number of these wolves, perhaps 80% by frequency of observation, appear to be single animals. The stragglers may represent either the walking wounded of the wolf population or the young in search of social group membership.

Caribou penetrations of the boreal forest have diminished in the last two decades; caribou have arrived at Stony Rapids only three times in the last 10 years. This is probably a result of the massive forest fires that have wreaked havoc upon the Stony Rapids–Wollaston Lake region for the last

7 years. But however far the caribou herds go for the winter, the wolves ultimately arrive in their vicinity, reaching densities of up to one per 6.9 square miles by late winter (Kuyt, 1972).

The wolf social system, as it appears from the yearly cycle, operates on a pattern of alternate concentration and dispersal of wolf packs to provide an adequate kill of caribou to meet the minimal functional requirements of the wolf population. The difference in pattern between wolf and Chipewyan is not as great as it may seem if we consider the physical differences between wolves and men.

THE WOLF AND THE CHIPEWYAN

The wolf and the Chipewyan, though occupants of different econiches, are competitors for the caribou. For both species, the caribou is the primary food source, yet neither species has been able to or has attempted to displace the other in past millenia. Even with the putative advantage of European weapons and technology the Chipewyan have made no headway in displacing the wolf. To appreciate the astonishing balance between these two predators, we must examine their physical differences to see how these affect them not only in the actual pursuit of caribou but also in the choice of strategies of subsistence.

Although wolf and Chipewyan are infinitely different in terms of their physical differences, they must perform functionally equivalent tasks to kill the caribou on which they subsist. But there are three major areas of difference: *(a)* the requirements for physical shelter, *(b)* the ability to travel over the ground and, *(c)* the capacity to utilize caribou for varying periods of time after an animal is killed. All these factors, as well as many others (for example, intelligence, technology, acuity of sensory systems) interact to provide each species with certain capabilities that are differentially utilized by the species as members of their respective social systems.

Requirements for Shelter

The wolf, because of its physical adaptations, is less tied to a physical shelter than is the Chipewyan. The only time the wolf seems to be tied to physical shelter is during the period that the pups are in the den, although it is probable that at certain times of the year (for example, at the height of the insect season and during strong winds on very cold days) the wolf must avoid certain areas unless shelter is available. Even in these circumstances the wolf can rapidly relocate to shelter by climbing a hill, digging a hole, or getting out of the wind.

The Chipewyan are in greater need of physical shelter than are the wolves, though the differences in their requirements are not as great as it might seem. Adult males are often caught on the trail or out hunting under circumstances that preclude their reaching previously established shelter. Under these circumstances the Chipewyan demonstrate a remarkable skill at finding and/or making shelter (Figure 4). Many families have spent winters in a canvas tent with no serious difficulties. As long as fire is available the Chipewyan are capable of surviving and functioning in their environment with little difficulty and few geographical limitations.

Although the Chipewyan have the capability of functioning in their environment with little concern for shelter they rarely exercise this capability. The Chipewyan choose to base their social unit from a fixed point that has a dwelling of some type as its center. The amount of time a dwelling location is used ranges from a single night to several months in

Figure 4. Chipewyan and caribou. The human need for shelter and the wolf's independence of this need provide a major contrast between the two species in their joint pursuit of caribou. (Drawing by John Slater.)

succession, and a single location may be used for several years though not for more than several months in succession (except in the recently established villages).

The Chipewyan cultural decision to use a fixed point as the location for a base is a function of the sexual division of labor, which allows the internal division of the Chipewyan hunting unit in subsistence activities. The wolf pack, apparently lacking a sexual division of labor throughout most of the yearly cycle, has no need of a fixed dwelling point for social reasons (or for physical ones) except at pup-rearing time. In both species the requirement for shelter is greatest during the process of rearing young but the differential life span of the two species makes a 3-month process for the wolves a multiyear process for the Chipewyan (Figure 5).

These differences in the requirements for shelter have significant consequences in the subsistence activities of the two species. The Chipewyan remain localized throughout their yearly cycle; no matter how often they change the location of their camp they always return to the central point.

Figure 5.   Mother and adolescent young. Though the pups require persistent care only for about the first 3 months of life, parent–offspring bonds persist; affection may be shown by play as well as by cooperation in hunting and in rearing new litters. (Photograph by Don Alan Hall.)

The wolf does this only during the denning season; the rest of the year the pack is free to move at will without having to return to a central point.

Mobility

Neither wolf nor Chipewyan seems to have a clear advantage in the area of shelter. In the second area, mobility, the wolf does have a clear advantage over the Chipewyan. The wolf is capable of covering more ground, faster and more thoroughly, than the Chipewyan, and mobility is the primary tactic of successful predation for both species. The disparity between the species in mobility is most apparent under marginal or difficult travel conditions. The Chipewyan are restricted in their movement at night; during freeze-up, breakup, and late spring; and when there is strong wind or extreme cold. The technological devices used by the Chipewyan (dog teams, snowshoes, boats, and so on) provide some increase in mobility but the increase is restricted by the necessity of caring for the equipment. The more elaborate the equipment the more its use restricts the freedom of movement of the Chipewyan, as the equipment itself must be maintained and preserved.

Caribou Consumption

Both wolf and Chipewyan can subsist on a diet consisting exclusively of caribou, but they have different methods of using other than freshly killed caribou. The wolf can use caribou that have been killed and left for periods of several months, and it can even work the remains of kills over a year old.

The Chipewyan lack the wolf's capacity to use old caribou—partly because of symbolic ideas about what is fit to eat—but they compensate with their ability to dry caribou meat. This gives an effective period of utilization of caribou of about 90 days during the summer and at least three-quarters of a year during colder weather. Drying meat serves two primary functions. First, it makes the meat portable so it can be easily transported and shared with other hunting units. Second, it concentrates food. It is possible to eat far more animal tissue in the form of dry meat than fresh. This helps greatly in the constant struggle for calories in cold weather.

Both species have a remarkable capacity to use caribou in forms far removed from fresh. The wolf's ability to gnaw on old bones is probably balanced by the Chipewyan ability to make soup out of old bones (and just about old anything), but neither species probably benefits particularly from consuming these items; they are a court of last resort, used only when a better nutritional input is lacking.

Tactics and Strategies of Caribou Hunting

Both wolf and Chipewyan kill caribou at almost all opportunities by whatever means possible, but the thousands of actual kills fall into a few patterns. The Chipewyan utilize three basic patterns, involving (a) watching and ambushing, (b) walking, and (c) using pounds and drives. By *watching and ambushing* I mean all variations of hunts that involve choosing a location from which to watch for caribou, moving to the caribou or to a point on their projected path once they are seen, and attacking. This strategy is the one most commonly used and is particularly favored during periods of high caribou density.

The second strategy, walking, is the simplest and most basic approach to hunting. It simply involves covering as much ground as possible in the hope of contacting caribou, at which point the appropriate tactics are applied. This strategy is sometimes a deliberate mechanism to kill caribou and sometimes incidental to other activities in which terrain is traversed. It is most favored during periods of low caribou density or difficult climatic conditions.

The third strategy, using pounds and drives, is now obsolete except for the occasional spearing of caribou while they are swimming. This basic strategy always involves the concentrating of caribou in a small area, usually through a created canalization of a known migration route, where the caribou may be killed in large numbers.

Wolves use the first two strategies in pursuit of caribou and some of the incidents described in the literature suggest that they may utilize the third strategy. The application of these strategies is different as a result of some of the physical and social differences between the species. The Chipewyan preference for the watch and ambush and pounds and drives in contrast to the wolf's preference for watch and ambush and walking derive from the Chipewyan practice of working from a central location at which the nonhunting part of the hunting unit is located and from the greater mobility of the wolf. The Chipewyan thus have a passive caribou hunting strategy whereas the wolves follow an active strategy. The Chipewyan disperse for reasons internal to their social system and they rely on the pattern of dispersal to bring the caribou to them. They then apply the strategies of hunting within a geographical region that is bounded by limitations derived from the time and ease of travel to and from the central point, a region selected for a number of reasons. A major factor is the Chipewyan estimate of the probability of the appearance of caribou. The Chipewyan goal is to take, in as short a period as possible, as large a number of caribou as possible.

For the wolf the primary determining factor of its social pattern and distribution is its vastly superior mobility. With the exception of the period when the wolf pack is tied to a den site, its social distribution is a balance between two separate and generally conflicting tendencies. The wolves must balance the inclination to remain in a fixed location until the caribou they have killed are utilized against the need to stay within striking distance of living caribou. The goal of wolf subsistence activity is the production of a sustained supply of freshly killed caribou throughout the year. This goal requires the wolf population to concentrate in the vicinity of the caribou and disperse around that point as a function of a social system that includes interpack tolerance.

The separation between the two species, human and wolf, is not absolute. The Chipewyan move their camps and males range out long distances; wolves hole up for varying periods and some wolves subsist on previous kills. The most obvious example of this occurs in the denning period of the wolf when, like the Chipewyan, wolves must fall back upon the utilization of caribou in other than freshly killed form.

In spite of these overlaps, the two species use characteristically different approaches to the obtaining of caribou even though they apply the same hunting strategies. Clearly, the greater capacity of the wolf to approach and close with the caribou compensates for the range of killing power of the rifle, bow, and spear. Though both species are locked into a struggle for utilization of the caribou it is not a direct struggle. Perhaps this point should not be stated in terms of a struggle for the caribou but from a viewpoint of complementarity. Both species must be viewed together as predators that bring pressure upon the caribou in a way that tends to keep a more uniform predation pressure upon the caribou than either species could manage alone.

Alternate Resources and Reciprocity

In one sense, both wolf and Chipewyan have made a commitment to caribou hunting that is ecologically inefficient. Though both species make use of many different sources of food neither species expends the energy to harvest secondary sources of food at a level near the potential these secondary sources could provide. Major secondary food sources that are underutilized are easier to document for the Chipewyan than for the wolf, but all the literature available indicates the wolf's specialization as a hunter of large game to the exclusion of systematic utilization of small game.

The Chipewyan underutilize as food resources moose, rabbit (snow-shoe hare and arctic hare, *ga* and *gacho*), grouse (especially spruce grouse), waterfowl (as an example, the Cree of the James Bay region have made goose hunting a successful substitution for the hunting of large game), and, most significant, fish. The fish population of the area repre-sents the largest and most consistent source of food available to the Chipewyan, yet fish are utilized only in the absence of caribou (or around the village, a contemporary but artificial situation) in other than the most casual manner. The Chipewyan frequently go into the bush for long periods without taking fish nets; they may discard them or lose them through neglect upon the arrival or anticipation of the arrival of the caribou.

This underutilization of resources, which may serve the temporal dura-tion of the entire socioenvironmental situation, is made possible by three factors: *(a)* the "willingness" of the Chipewyan cultural system to exist at a level well below the absolute carrying capacity of the environment (a feature of most nonagricultural systems), *(b)* the complex cultural prac-tices around the drying of caribou meat, and *(c)* the reciprocity network provided by the Chipewyan kinship system.[4]

The Chipewyan system has opted for a pattern of simultaneous disper-sal and concentration of hunting units to localized geographical areas to provide a maximum kill of caribou, with subsequent drying of caribou meat for redistribution through the reciprocity network—which may in-volve physical relocation of hunting units—to provide for the needs of the population. The wolf system has opted for a pattern of alternate concen-tration and dispersal of packs in a matrix fixed to other wolf social units rather than geographical features; lacking a reciprocity system for pre-pared meat, the wolf system relies upon the continuous taking of caribou and the free[5] utilization of excess caribou kills.

CONCLUDING REMARKS

In this chapter I have attempted an ethnographic comparison of the subsistence systems of two different but remarkably similar cultures. I

---

[4] It is not feasible to explain the anthropological fascination with reciprocity in the length allowed for a footnote, and I refer the reader to Mauss's (1954) brilliant little book, *The Gift*. For the reader who is not a social scientist I should like to stress that it is not the distribution of the dry meat or the dry meat itself that matters; what is significant is that the act of giving (regardless of what is given) creates a patterned network among the exchanging parties that becomes a mandatory framework for future interaction.

[5] I presume it to be free, though it would support my argument more if it were patterned in some way.

have, I hope, demonstrated that Chipewyan and wolf are part of a single predator–prey relationship with the caribou, in which the same fundamental strategies, modified by the physical capabilities of each species, are used in a complementary manner by each. This relationship is culturally conditioned and selected from the possibilities allowed by the environment. Though comparison of social structure is beyond the scope of this chapter, I have indicated some of the similarities derived from the two species' pursuit of a common occupation. I hope that this analysis will serve as a stimulus to further investigation of the sociology of the wolf and the investigation of other animal hunters as a source of models for our own evolutionary antecedents and their development of social hunting.

REFERENCES

Bone, R., Shannon, E., & Raby, S. *The Chipewyan of the Stony Rapids region.* Mawdsley Memoir No. 1, Institute for Northern Studies, University of Saskatchewan, Saskatoon, 1973.

Helm, J., & Damas, D. The contact traditional all-native community of the Canadian north: The upper Mackenzie "Bush" Athapaskans and the Igluligmiut. *Anthropologia,* 1963, *5,* 9–22.

Joslin, P. W. B. *Summer activities of two timber wolf* (Canis lupis) *packs in Algonquin Park.* Unpublished master's thesis, University of Toronto, 1966.

Kuyt, E. *Food habits of wolves on barren-ground caribou range.* Report Series No. 21. Ottawa: Canadian Wildlife Service, 1972.

Mauss, M. S. *The gift: Forms and functions of exchange in archaic societies.* New York: Free Press, 1954.

Mech, L. D. *The wolf: The ecology and behavior of an endangered species.* Garden City, New York: Natural History Press, 1970.

Mech, L. D. Wolf numbers in the Superior National Forest of Minnesota. U.S. Department of Agriculture Forest Service Research Paper NC-97. Saint Paul, Minnesota: North Central Forest Experiment Station, 1973.

Parker, G. R. Distribution and densities of wolves within barren-ground caribou range in northern mainland Canada. *Journal of Mammalogy,* 1973, *54*(2), 341–348.

Rausch, R. A. Some aspects of the population ecology of wolves. *American Zoologist,* 1967, *7,* 253–265.

Sharp, H. S. *The kinship system of the Black Lake Chipewyan.* Unpublished doctoral dissertation, Duke University, 1973.

Sharp, H. S. Introducing the sororate to a northern Saskatchewan Chipewyan village. *Ethnology,* 1975, *14*(1), 71–82. (a)

Sharp, H. S. Trapping and welfare: The economics of trapping in a northern Saskatchewan Chipewyan village. *Anthropologica,* 1975, *17*(1), 29–44. (b)

Part II

COMMUNICATION AND COGNITION

An attempt to understand protohuman cognition and communication systems begins with a study of the anatomy and physiology of human thought and language, and leads invariably to a study of animal cognition and communication. Unfortunately, certain barriers to communication exist between man and animal. Some of the barriers are real in the sense that all animals, including humans, are anatomically and behaviorally equipped to communicate most easily with members of their own species. Other barriers to understanding the communication of animals are special to our own species in that we have *chosen* to distinguish, explicitly as well as implicitly, and ethically as well as perceptually, between ourselves and other animals.

Clearly a large portion of the barrier between animal and man lies within human perception, particularly the perceptions we have concerning language and thought and the role each plays in human communication. Technically, communication is defined as any act performed by one individual that causes a change in the behavior of another individual—by this definition almost all animals are communicators. Human language, an explicit symbolic system, overlaps the province of communication but also differs from it in several ways. Though language sometimes functions to facilitate communication, it also functions to facilitate thought and is not always used as a communication medium. Consider, for example, an individual sitting in his room, engrossed in thought; this person is not communicating! It is even more important to note that most of human communication is not conducted by language but by nonverbal means and, to a large extent, by means we ourselves are not conscious of.

Furthermore, language itself is often used nonconsciously and is used more often to establish and maintain social relationships between communicators than to transfer information. Much of human nonverbal communication must be considered akin to communication systems used by other animal populations. Indeed, we may even share specific signs with various animal groups, particularly with the primates, our closest phylogenetic relatives. Nonverbal communication in *Homo sapiens* has been the object of many studies under the headings of paralinguistics, body language, and kinesics (Birdwhistell, 1970; Davis, 1971; Fast, 1970).

Clearly, one of the links between animal and human communication exists in our mutual dependence on body language, gesture, vocal pitch, and facial expression as indicators of feelings or intent. Yet another bridge may exist between animal and human communication and this we can term *social interaction communication*. This is so new an area of study that as yet its boundaries are diffuse, but it can be defined tentatively as the transfer of specific information from animal to animal by social interaction activities (facial expressions, gestures, grooming behaviors, and the like). It is necessary to develop a new category for this communication, which stands in opposition to the long-held assumption that the content of animal communication is entirely emotional and immediate, that is, conveys a feeling or desire that an animal has at one particular time but does not convey information about the environment (Lancaster, 1968). Though not yet conclusive, evidence that this may not be true is suggestive and will be discussed briefly.

E. W. Menzel, Jr. (1971) carried out a series of experiments with eight juvenile chimpanzees in which he tested their ability to communicate specific information. These chimps were all wild-born animals but at the time of the experiments they had been living together in a field cage for a year. The experiments involved hiding a number of objects, including some desired objects such as bananas and some "fear objects" such as a simulated snake. The objects were hidden in the field cage while the chimps were removed. One member of the group was shown the location of the objects and was then returned to its cage. Soon all the young chimpanzees were released and investigators recorded the procedures that the chimps used to locate the test objects. After more than 1000 test trials utilizing all individuals as leaders, it was clear that the leader—whoever he or she happened to be in specific instances—communicated to the other animals not only that he or she knew the location of the object, but what kind of item it was. Control experiments in which no individual was shown the location of the cached objects established definitely that the behavior of the animals was clearly keyed to the communication of specific information by leaders when prior information was available.

What kind of communication medium did the chimpanzees use? It was clear that they did not use vocalizations or a system of gestures. Menzel infers that the leader transmitted subtle indications of his or her directions and of the goal he or she expected to find. Within the context of a small, intimate group, individuals were able to perceive intention and even explicit information from minute interactions and patterns of behavior—glances, shrugs, styles of gait, and so on. Menzel suggests that by this means chimpanzees communicate information without a symbolic code such as that used in human language systems.

As an impediment to understanding the communication systems of animals, our preoccupation with systems that *appear* similar to our vocal language is second only to our preoccupation with the uniqueness of human language. The songs of birds, for instance, were among the first communication systems studied by ethologists; and the howling of the wolf and singing of the coyote have received more serious study than have their scent-markings, which may be of equal or greater importance. Of course a part of our handicap in dealing adequately with scent-marking is physical; our sense of smell is simply inadequate to perceive and decode the wealth of information contained in scents. Similarly, early attempts to teach language to chimpanzees were made using human vocal symbols. These attempts were abysmal failures. In contrast, recent experiments in teaching chimpanzees sign language have been extremely successful. Not only have chimps mastered a large number of symbols, but they have also invented new ones, and most important they have put symbols together in grammatically meaningful sentences.

The important question, as yet unanswered, is this: If chimps *can* use language, why do they apparently not use it in their natural setting? To answer this question we can offer several alternative hypotheses:

1. No chimp has invented language; hence, for chimps to use language they had to be exposed to it in another species. As an example and possible analog from human behavior we may suggest that *Homo sapiens* for many thousands of years had the ability to use written language, but writing came into gen-

eral use only during the last 5000 years, because the circumstances promoting its invention came into being then.

2. Chimpanzees can use language and it may have been invented on several occasions, but in their natural habitat and social environment language use is not an adaptive trait and hence is not practiced.

3. Chimpanzees are using a form of language in their natural habitat but the manner in which they communicate specific information—the channel that they use—is so different from vocal–verbal language that it has escaped our notice. In support of this possibility, observers have described instances in which several chimpanzees have cooperated in deceiving a hunted animal (see Teleki, 1975). Another indication that chimpanzees habitually use a very different but perhaps more efficient channel for communication comes from studies of groups of chimps who have learned sign language. When their human trainer is not present, they do not communicate by sign but revert to natural chimp channels.

It is possible and perhaps even likely that some form of explicit symbolic communication is used to coordinate chimpanzee hunts. Of course, the chimpanzees' explicit communication system may not have as broad or as general a scope as that used by our species. It may depend, for instance, on intimate knowledge of the other animal's behavioral cues, and may not generalize to entire groups as human languages do.

Similarly, hunting strategies used by wolves may be based on an explicit form of communication in which facts about the environment are shared and actions are planned and coordinated (Figure 1). If this can be demonstrated, the communication system would have to be judged protolanguage, if not language. Though human language is specific to humans, the functions of communication performed by that system are performed adequately by language systems specific to other animal groups.

EVOLUTION OF THE BRAIN AND CONSCIOUSNESS

Anthropologists, philosophers, and psychologists have long concerned themselves with the problem of the human brain: How can its evolution be explained? As evolutionists we acknowledge that so large and energy-expensive a structure must confer a tremendous

Figure 1. Expressive face. Highly social species, such as the wolf and most primate species, tend to have very expressive faces for the communication of subtle moods and perhaps for the explicit exchange of information. (Photograph by Don Alan Hall.)

adaptive advantage, else natural selection would not have promoted its development. For instance, the brain comprises only about 2% of the body's weight but it uses 20% of the blood supply—this is but one indication that physiologically it is an expensive structure. Furthermore, the human brain has required correlated anatomical adaptation, notably in the region of the pelvis. In man, the pelvis has adapted to become a basin-shaped structure whose wide and curving surfaces serve as muscle-attachment areas that permit us to stand, as well as run, on two legs (clearly an engineering feat!). But the pelvis also must serve as a support for the inner organs and for the development of a human child, and must still have an opening large enough to permit the birth of a large-brained infant. That adaptations in the pelvis have been accomplished to accommodate brain size indicates further that the human brain must have made a significant contribution to the species as a whole. What *are* these contributions?

By focusing on the recent expansion of technology and science, some writers have argued that the brain's potential has only recently been tapped and hence have implied that its evolution cannot be explained. Arthur Koestler (1967) presents this point of view eloquently in his "Parable of the Unsolicited Gift" in the book, *The Ghost in the Machine*. Koestler tells the story of an Arab merchant, Ali, who prayed to Allah for an abacus so that his customers would not take advantage of him, as they had been doing. Instead of an abacus, however, a giant computer was presented to the merchant—but, alas, it came with no instructions. By persisting with it, Ali found that if he kicked and beat on the machine in certain patterned ways it would do the sums he required. He prospered, and passed the machine on to his heirs who continued to manipulate it in the same way for generation after generation. As Koestler presents it, we are Ali's heirs and the human brain is the unsolicited gift that we are only now beginning to learn to appreciate.

Offering a contrasting interpretation of the origin of the human brain, anthropologists have often suggested that the brain's function among prehistoric peoples, though slightly different from its function at present, was equally critical and equally potent as a focus for selective force. One often-suggested function of the brain in nonliterate societies is to store information, both specific environmental data and the kinship structures and specific kin relationships that are crucial to the social organization of most nonindustrial peoples. The great wealth of detail regarding relatives, both dead and living, and the environmental information that is possessed by these peoples require a "storehouse." The evolution of intricate social forms, as well as the requirement for memory, must have selected for increase in brain size, according to this view. Furthermore, it has been suggested that initial selection of a large memory capacity, superior to that of other members of the primate order, came when early hominids began a persistence hunting way of life. Then survival depended on intimate knowledge of a specific environment coupled with the ability to share information with other members of the social group.

Primate anatomists have emphasized that the supreme sensory ability that marks the primate order is vision. Originally selected to enhance hand–eye coordination among primates living in trees, vision served well the emergent hominid. On the open savannah the problem was to locate and obtain food, and to spot and remember the location of hazards and food sources. Keen vision, memory, and explicit communication skills have all been judged as contributors to selection for a brain in which visual imagery plays a critical role.

Though the traditional anthropological view summarized in the preceding two paragraphs provides a frame of reference for explaining the evolution of the human brain, it provides only the skeleton of that framework. Discoveries in diverse fields—including computer science, psychiatry, and neuroanatomy—are adding new dimensions and more detail to our interpretation of the evolution of function in the human brain. Rather than tearing down earlier edifices, however, these new discoveries and insights are refining and making more explicit, and hence more testable, previous concepts. The following model of brain function is viewed as a supplementary, tentative model.

Simply stated, this model postulates that the brain acts as an information processor; it stores information, sorts information, and decodes and recodes information. The brain performs these functions whether we will it or not—it does this continually, in *Homo sapiens* and in other animals. One of the most interesting phases of brain activity occurs during the period of sleep in which dreams occur, the period known as rapid eye movement (REM) sleep. We have no explicit way of determining whether animals dream (i.e., experience visual imagery) in the same way that humans do, but it has been established that they experience the same physiological behaviors (e.g., the same brain-wave patterns) that humans do when they dream. Hence, most researchers infer that animals also experience visual imagery, or dreams. Since the human brain differs anatomically from the brains of other animals in quantity and proportion rather than in kind, we should expect that the same basic processes occur from species to species. From an anatomical point of view, the most obvious distinguishing feature of the human brain is its size, particularly the size of the cortex, or outermost layer. The human brain is unique in the amount of associative areas that connect various sensory modalities. More directly, the layman notices that the human brain has an additional burden put upon it: consciousness. The human brain must organize stored sensory experience in such a way as to permit consciousness and to make consciousness intelligible.

During REM sleep, when dreaming occurs, the electrical activity of the brain is similar to its patterns during waking; all the signs indicate an alert, waking brain even though muscle tone and other signs indicate sleep. The discrepancy between what appears to be a waking brain and a deeply sleeping body has prompted some experts to refer to the REM period as *paradoxical sleep*.

Brain function in this period is especially intriguing. During this phase the brain is apparently rearranging its stored data and integrating data taken in during the past day with data stored previously. More fundamentally, it is revising old models of reality to conform with newly obtained sensory information and experience (Allison & Twyver, 1970; Laughlin & D'Aquili, 1974).

Clearly, all animals face the same problems concerned with the recording, processing, storing, and retrieving of information; all brains must process data by which to guide behavior. Consider the gibbon brachiating from branch to branch. Certainly the gibbon possesses an innate ability to brachiate. A portion of its expressed ability includes visual perception, the ability to judge distances, and hand–eye coordination. Learning also plays a role in gibbon locomotion, for the gibbon must continually assimilate data concerning its immediate environment. Its next movement must be based on what it has learned about features of its environment in the past as well as upon the visual picture of the immediate environment. It must be able to estimate the strength of various branches, for example, and it must be able to estimate its own power to propel itself. We do not know what the gibbon is conscious of as it makes its way through the trees— probably most of the neural processing required to carry it safely among the branches is not at the conscious level.

Compare this picture with that of a person driving a car. Clearly the driver has placed himself in a position as precarious as that of the gibbon. Information about the immediate environment and the environment into which he is propelling himself must be gathered, processed, and analyzed against the background of former experience. The difference between the man in the car and the gibbon in the trees is that at least some of the experience of the man driving the car is assimilated vicariously and consciously. The power of consciousness lies in the ability to direct and program that vast computer, the (subconscious) brain. In this way a person can choose to learn to drive a car and by individual study and conversation with other persons can benefit vicariously from experiences of others; hundreds of hours of experience can be assimilated in a very short time.

We are taught the rules of driving and by study and practice we assimilate them. We are even taught, explicitly, about driving conditions that we ourselves may never experience or may experience very infrequently; these facts are stored in a "neural bank" and are actively available to the neural system should they be required, the speed of the retrieval being phenomenally quick. Though we may believe that our conscious mind is in control of our driving it is certain that much of the control is exerted at a subconscious level. Clearly, a large part of the brain's function—and perhaps a large portion of its absolute mass as well—must involve *determination of which portions of the array of stored data surface to the conscious mind.* In this view, one of the most important functions of the human brain is to act as a "traffic cop" controlling the flow of stored information, determining priorities and access rights to the individual's zone of attention or consciousness.

In fact the conscious mind *and the structures that support it* must be truly revolutionary products, even though the main power of consciousness lies in its ability to direct, meaningfully, structures that were already present. To bring data into the conscious mind basically means to make a thing of an idea, a reflection, or a mental process; it entails giving to a fleshless and insubstantial "thought" the characteristics of an object.

Language facilitates the process. Though language is intimately bound up with consciousness it does not appear to be associated directly with vital processes of thought that are not conscious. This conclusion emerges from studies of the different functions of the two sides of the brain. Since these studies have been discussed in detail in many other publications they will only be summarized here (Dimond & Beaumont, 1974).

For years neuroanatomists have known that the two sides of the brain differ functionally and that in most persons one side (usually the left) is heavier, slightly larger, and in some more subtle respects, dominant. Dominance appears to be inversely related to handedness; right-handedness is usually associated with cerebral dominance of the left hemisphere. What is cerebral dominance? One side of the brain has the principal role in language acquisition and in the initiation and control of voluntary motor activity. Essentially, this means that the dominant hemisphere has the leading role in controlling and directing consciousness. (*Dominance* is a misleading term in that it implies superiority or greater strength, which is not properly a part of the division of labor between halves of the brain. For this reason many investigators prefer to designate the functional distinctions of the two sides of the brain as "lateral asymmetry" rather than as dominance.)

Physiologically, dominance can be broken by severing the corpus callosum, a bundle of nerve fibers that links the two sides of the brain. This operation has been performed experimentally with monkeys and has also been used successfully, as a last resort, in treating epilepsy. Perception and linguistic tests made on persons with a resulting "split brain" show that, truly, in these people the left brain does not know what the right brain is doing. Each hemisphere has its individual identity and when the normal mechanism that integrates their activities is violated we can obtain a representation of the functional activity of each. The picture that emerges is that only one portion of the brain is thoroughly verbal; the nondominant hemisphere (usually the right) is capable of responding to verbal commands but not of expressing its activities verbally. It must be emphasized that the dominance concept does not mean that activities of the nondominant side are nonessential or nonfunctional; it means only that they are not involved in the process of conscious thought and language expression.

The dominant hemisphere is concerned with verbal, mathematical, and symbolic knowledge; the nondominant hemisphere excels at visuospatial tasks. These general principles have been tested on persons with a split brain in the following way: One eye is masked and a command is presented through the unmasked eye. This command is carried to the brain hemisphere opposite to the eye exposed—for example, if the left eye is exposed the command goes to the right brain. Or a puzzle problem may be presented, again, to only one hemisphere. Many times a problem that requires mechanical aptitude is solved very neatly by the individual's right hemisphere—but the person cannot describe what he has done. The behavior may be entirely appropriate but it is apparently nonverbal, and, as we normally understand the concept, nonconscious as well.[1]

Nonconscious thought seems a contradiction in terms but we, along with other animals, practice it and in fact our lives depend upon it to a larger extent than upon conscious thought. Psychiatrists since Freud have argued that the functioning of the nonconscious mind affects behavior. Now neuroanatomists and physicians have found that we can use our conscious minds to intrude into the domain of supposed autonomic functioning. For example, patients who suffer from migraine headaches can be taught to control the flow of blood to the head so as to return the body to equilibrium and in effect control the cause of migraine. The field of study involved with con-

---

[1] Earlier it was noted that the *Homo sapiens* brain is anatomically novel in its association areas that link sensory modalities. This means, for example, that objects that we perceive by vision have associated smells and textures within our mental construct of those objects. We have learned to articulate mechanical processes; however, a brief illustration will indicate that our skill in this area is still rudimentary, even in persons who have a unified brain! Have you ever tried to tell someone, without illustration or mechanical aid, how to load even the simplest camera, or how to thread a sewing machine? Clearly, "a picture is worth 1000 words"—our ability to verbalize mechanical instructions (that is, to associate the activities of the many separate compartments of our two brains) is as yet far from perfect. This example reemphasizes the point that the human brain is different from other brains in the degree of particular development; we do not have brains representing the "quintessence of development."

scious control of the automatic areas of the body is termed *biofeed-back;* this refers to the technique of control (Brown, 1974). Persons are provided with an apparatus to measure a particular bodily process and then are asked to alter that response. Hence, they know at once whether their thought process is achieving the desired goal, and from this process comes the term *feedback*.

Although the practice seems novel or even revolutionary, the learning procedure is not new; its only novel aspect is that it is being applied to a new sphere of activity. Consider, for example, the case of the novice learning to play basketball. If he did not know whether or not his tosses pushed the ball through the basket, would he stand a good chance of improving his aim? Not very likely. Most learning and teaching is in fact based on a feedback system. Clearly, we do not know which muscles and nerves we manipulate in order to improve our skills. In a sense we are like the chimpanzees who are just learning to manipulate sign language; we are extending our conscious control into a new area of mental and physical process. Both conscious and nonconscious brain functions are at work maintaining our complex brain–body system; the problem we face is one of attaining a harmonious balance including judicious use of the potentials of both systems.

EXPLORING THE EVOLUTION OF CONSCIOUSNESS

We have tried to show that unconscious mental processing in humans and in animals is most critical to survival; that is, it provides the appropriate behavioral response to the environment. The power of conscious thought lies in the ability to treat an idea or a reflection as if it were a thing and to call into play the tremendous reservoirs of mental process that lie at the subconscious level. In practice, there is a gradation between conscious and nonconscious, and there also exists a constant interaction between them. Given these premises, what model can be developed to account for the evolution of consciousness? To what extent does consciousness exist among other animals? Is it expressed in their sociocultural life and in their communication systems? In our species, consciousness is entwined with language, which is dependent on the vocal–auditory channel—can the capacity evolve without that channel? Jane Hill (1972) believes that hominids were preadapted for vocal language by virtue of their primate heritage and that freeing of the hands for tool use, together with upright posture, provided the final anatomical impetus for its evolution. Clearly other animal populations, with different pre-adaptations, would evolve language by other channels.

Though the human capacity to use language and to manipulate and communicate ideas consciously and explicitly distinguishes our species from other animal groups (Hill, 1972), most of our behavior, thought, and communication occurs at the nonconscious level but is nonetheless exceedingly complex and subtle. Furthermore, just as a person waking from a dream appears on a bridge between two zones of reality the nonconscious and conscious processes grade into each other. Similarly tests of persons with their brain hemispheres disjoined indicate that the nondominant hemisphere, thought incapable of instigating speech can still digest verbal data and can make appropriate and complex responses to it; hence, these mute hemispheres yet possess a kind of consciousness, perhaps akin to that possessed by some animal species (Zangwill 1974). In order to understand the phylogenetic development of consciousness and explicit communication we need to search in other animal groups for activ ties and mental processes that indicate abilities of this kind, even if rudimentary in form Clearly, chimpanzees that have been taught to use sign language possess this ability. Do wolves also possess it? Since their anatomical endowments differ more from our own than do those of the chimpanzee it is more difficult to devise an ideal between-species communication medium Consequently we must study their natural behavior to determine whether they are communicating symbolically—to see whether, even in rudimentary form, they are communicating specific information and are manipulating ideas as if they were objects The chapters that follow treat three aspects of wolf cognition and communication.

REFERENCES

Allison T. & Twyver H. an The evolution of sleep *Natural History*, 1970, 79 (2), 56–65.

Birdwhistel, R *Kinesics and co text*. Philadelphia Univ of Pennsylvania Press, 1970.

Brown B R *New m nd new body* New York: Harper, 1974.

D vis F *Inside intu tion what we kn w about non-verbal communication*. New York McGraw-Hill, 1971.

Dimond, S J & Beaumont, G. (Eds.). *Hemisphere function in the human brain* New York Wiley 1974

Fast J *Body language*. New York. Evans, 1970.

Hill J On the evolutionary foundations of language. *American Anthropologist*, 1972 74, 308–317

Koestler, A. *The ghost in the machine*. New York. Macmillan 1967

Lancaster, J. Primate communication systems and the emergence of human language. In P. C. Jay (Ed.), *Primates*. New York: Holt, 1968.

Laughlin, C. D., Jr., & D'Aquili, E. *Biogenetic structuralism*. New York: Columbia University Press, 1974.

Menzel, E. W., Jr. Communication about the environment in a group of young chimpanzees. *Folia Primatologica*, 1971, *15*, 220–232.

Teleki, G. Primate subsistence patterns: Collector–predators and gatherer–hunters. *Journal of Human Evolution*, 1975, *4*, 125–184.

Zangwill, O. Consciousness and the cerebral hemispheres. In S. J. Dimond & G. Beaumont (Eds.), *Hemisphere function in the human brain*. New York: Wiley, 1974.

5

Communication, Cognitive Mapping, and Strategy in Wolves and Hominids

Roger Peters

The effect of variation in ecology on social behavior in several species of primates and carnivores has been well documented. For this reason, Schaller and Lowther (1969), Kortlandt (1965), and others have suggested that the study of wolves and other social carnivores might provide insight into the problems and opportunities of socially hunting hominids. This view assumes that hunting was an important part of hominid ecology for significant periods of time.

Meat accounts for a relatively small proportion of energy and protein intake of contemporary hunter–gatherers. These proportions, however, may not reflect the selective significance of hunting, which might be decisive for survival in restricted periods or regions, and which is highly valued in many cultures and could thus contribute to fitness far more than dietary measures suggest.

Without addressing the issue of the extent to which hunting large animals in groups may have been a selective force in human evolution, this chapter extrapolates from generalizations about the intellectual adaptations of wolves to hypotheses about analogous adaptations in our ancestors. The central assumption in this extrapolation is that big-game hunting requires the use of a large area. DeVore and Washburn (1963) wrote the following:

> In the evolution of human behavior, hunting is the best clue to the size of the range. . . . The major difference between the baboon range and that of human hunters is the vastly larger area which humans, like the other large carnivores, must control [p. 353].

Density of prey is an important determinant of the size of a hunter's range. Although the prey of hominids may have been more densely distributed than that of wolves, the ranges of social carnivores in East Africa, where prey is much denser than it is in Minnesota, also are large and comparable in size to those of wolves (hundreds of square kilometers). The best data on the size of the range of wolf packs were obtained by Mech (see Chapter 7).

Another indication that the ranges of hominids might have been comparable to those of wolves is the similarity in speeds of long-distance travel in wolves and in the early hominids' human descendants. Both wolves and men are known to have covered distances of about 160 kilometers in a 24-hour period. Therefore, although earlier hominids may not have been as fast as modern men, the scale of the hunting areas of the two species probably was roughly comparable, and considerably larger than those of nonhuman primates. Baboons and chimpanzees, for example, have ranges of less than 35 square kilometers, and rarely travel more than a few kilometers at a time.

In hominids, as in wolves, the behavioral solutions to the problems of hunting in a large area probably included (a) a communication system that created an integrated group but also allowed members to separate and rejoin, and (b) cognitive maps, which allowed efficient and insightful travel. Wolves, unlike hominids, developed these solutions with the aid of powerful odors, sensitive olfaction, and natural weapons. Hominids, without these abilities but with greater behavioral plasticity and more cortex, would have had to develop visual, vocal, and intellectual means of adapting to a wolflike economy.

COMMUNICATION IN WOLVES AND HOMINIDS

Hominids probably found vocal and visual ways of communicating the kinds of messages wolves transmit by means of odors. The functions of olfactory communication among wolves almost certainly include recognition of individuals and the expression of moods, especially excitement, submission, fear, aggression, and solidarity.

Among hominids, vocalization may have provided an effective means of recognizing individuals during presumably frequent rejoinings of group members, particularly in dense cover or after dark. Natural selection may have favored the ability of individuals to produce distinctive patterns of sounds, just as it has favored the ability of individual wolves to produce distinctive odors. Distinctive voices could produce distinctive vocalizations, thus creating the forerunners of names. The distinctiveness of human voices. which today allows recognition of individuals, even over the telephone may thus be an adaptation to a way of life that placed a premium on quick identification of fellow hunters.

Much social sniffing among wolves serves as an alternative to, and possibly as an inhibitor of, fighting. Hominids may not have found each other's smells sufficiently distracting and could have increased their personal appeal by costume gesture, or vocalization. In *Steps to an Ecology of Mind*, Gregory Bateson (1972) suggested that the major function of human social vocalization is the reassurance that locutors gain by continued vocal production, and cites as evidence the common feeling of disturbance that in many cultures results from silences between men who do not know each other well. Presumably the referential function of language would have been even less apparent in the chatter of hominids than in that of modern humans. In this view, words are a relatively recent refinement of the ancient practice of vocalizing a lot to say very little.

Wolves often use olfactory investigation and high-pitched vocalizations to express affiliation, and sometimes such expressions culminate in a group ceremony: an excited milling about, with barks, whoops, growls, facial sniffs, and playful pawings that in turn sometimes lead to a group howl. Human vocalizations in the form of sweet nothings and song may represent the modern equivalent of early hominid expressions of togetherness. In any case, hominids who had to hunt together, sensitive to one another's every move, may well have developed analogs to the group ceremony, and sound seems to be the most likely medium for this expression. Like odor, its physical properties are well adapted to the formation of a group production in which individual identities blend yet remain distinct.

The fact that hominids lacked wolves' odors and olfactory sensitivity does not imply that olfaction was unimportant in hominid communication. In fact, many of the forms of direct olfactory communication in wolves, especially those associated with reproduction, have homologs among other mammals, including humans It is therefore likely that these forms of communication occurred among hominids, too. Olfaction (and gustation) may have been important in establishing bonds between hominids, just as they are in humans (Kinsey, Pomeroy, & Martin, 1948).

In particular, it has been demonstrated that the short-chain fatty acids that account for much of the anogenital odor of wolves act as olfactory sexual stimulants and attractants in rhesus macaques *(Macaca mulatta)* (Michael, Keverne, & Bonsail 1971). If, as my informal experiments suggest, some of these substances (especially propionic acid) have pheromonal properties in wolves as well, the use of the same chemicals by animals as different as wolves and monkeys indicates that they may be found in other mammals, and may have been used by hominids.

COGNITIVE MAPS OF WOLVES AND HOMINIDS

The term *cognitive-like map* was coined in 1948 by Edward Tolman. He used it to refer to a mental representation of the location of a goal and paths that allowed ''insightful'' travel, that is, travel by a shorter, often novel route, rather than by a more familiar route. In a classic but imperfectly controlled set of experiments (in rats, as in wolves, olfactory cues from previous travel can provide clues that might facilitate route choice without knowledge of spatial relationships), Tolman and his students showed that after rats had become familiar with a maze of elevated runways they readily chose novel shortcuts; they chose a longer, unblocked route after encountering an obstacle that blocked two preferred and shorter routes; and in general their choices were consistent with the hypothesis that rats remember spatial relationships rather than simple stimulus–response connections.

The term *cognitive map* has been broadened, to refer to any conception of a geographical area, and refined by Kaplan (1973) to refer to a neurophysiologically embedded associative structure of nodes and paths. In this chapter *cognitive map* will be used in Tolman's original sense, and in deference to recent work by Kaplan the concept will also provide a basis for discussing the planning of actions not directly related to terrain.

Kaplan views cognitive maps as networks of representations of key locations (nodes) connected by paths of association. (A representation is a mental model of some portion of the environment and is itself a network of associated features.) Kaplan's (1975) discussion of nodes, that is, key places that are parts of several different paths (choice points) is particularly interesting in terms of the behavior of wolves:

> Making one's way, that is, traversing a path in space from start to goal, requires first of all that one recognize critical loci where choices are to be made. . . . Landmarks play a central role in the identification of both choice points and direction. The more distinctive the place, the more readily it serves as a basis for decision making and a change of direction . . . whether the distinctiveness is a blunt sensory contrast or the outcome of a refined sensitivity to the character of the surrounding area [p. 108].

There are several kinds of evidence that not only suggest that wolves use cognitive maps but, in conjunction with Kaplan's theory, provide some indications about what the elements of these cognitive maps might be. First, there is evidence that suggests that wolves have both the opportunity to acquire cognitive maps and the occasion to use them. The second kind of evidence indicates that wolf travel is often insightful, and thus implies that wolves have some kind of mental representation that allows them to plan their routes. The third kind of evidence comes from the features of the environment that wolves mark, features that must be recognized if wolves are to make informed decisions about long-distance travel.

A cognitive map is useful to any animal that must travel to find food or other resources in an area that is large relative to the animal's scanning capabilities but restricted enough to permit repeated use of the same terrain. Several investigators have inferred the use of cognitive maps from behavior in the field or laboratory of their subjects (Kaplan, 1973; Leyhausen, 1965; Tolman, 1948).

An efficient means of remembering locations is particularly useful to social carnivores for several reasons. The large animals they eat are less densely distributed than the small animals and vegetation that nourish other creatures. Social carnivores must therefore travel widely in order to encounter and test as many potential prey as possible. Furthermore, the distribution of their prey is often clumped, and this factor places a premium on knowing where to look. Finally, social carnivores must often find not only prey but also their way back to a den or rendezvous site in order to feed their young.

Cognitive maps are of course not the only means of finding one's way. It is conceivable that an animal might learn all the paths it needs by rote (response learning), remembering an unorganized set of "strip maps." Or, like an insectivore, it could confine its movements to trails it previously marked, or those marked by fellow pack members. Both these alternative strategies would restrict exploration and place the animal at the mercy of weather or other natural forces which could obliterate or alter the cues needed for choice of route. Neither alternative allows for much flexibility or such "insightful" behavior as taking a shortcut or detour, which depends on knowledge of alternative routes and spatial relationships.

Cognitive maps are *useful* only when travel has a destination. Since hunting is the major function of travel by wolves (Mech, 1970), regions where prey is likely to be found, places where kills have been made, and routes connecting them are likely destinations. Since wolves must return to dens and rendezvous sites during the summer to feed their young, those locations are also likely to provide goals. These destinations may offer some clues about the elements of the cognitive maps of wolves, elements that probably include locations (of dens, rendezvous sites, and kills), regions (or prey concentrations) and routes (connecting these locations and regions).

The best evidence for the use of cognitive maps by wolves is the same kind used by Tolman years ago, namely use of multiple routes, and especially shortcuts. In my field study of wolf travel (Peters, 1974), shortcuts were defined as excursions (from a road, trail, or waterway) that *(a)* had been followed for at least a kilometer, *(b)* returned to the route or another connected with it in such a way as to reduce the distance or effort of travel, *(c)* were followed by use of the continuation of the route for at least a kilometer, and *(d)* occurred under conditions that precluded auditory contact with a wolf who had not taken the shortcut.

Shortcuts imply that wolves use mental representations of routes that contain geometric relationships (Figure 1). The major differences between their travel and travel based on a long-distance sense of orientation is that wolves use established paths and choice points rather than, like some birds, traveling in straight lines or relying on distant or pervasive physical configurations (Adler, 1970).

The adults of a pack that I studied in the early 1970s in Minnesota seemed to grow more familiar with their territory during the three winters that I followed them. In later years they consistently cut corners that they formerly traced. Their repeated use of shortcuts thus seems to have increased with their familiarity with routes. Pups never made a shortcut except when accompanied by their parents. These observations support the notion that wolves learn the routes and locations of their territories by experience with them and that these routes and locations are represented mentally in a way that allows insightful planning of travel.

Additional evidence about the elements of wolves' cognitive maps comes from their marks and eliminations. The parts of the environment to which wolves apply the odors of their tracks, marks, and eliminations become vivid olfactory entities, set off from their unanointed surroundings by complex and fascinating odors. These vivid entities are the features most likely to be perceived, remembered, and included in cognitive maps. Since wolves place sign on their paths every 250 meters on the average, a wolf traveling on a route he has used before is rarely more than 125 meters from a relatively powerful source of familiar odor. These odors occur at intervals short enough so that they, along with the odor from the wolf's feet, and scent marks from previous trips, make the path an olfactory unit as well as a visual one.

Since scent marks are not only distributed along paths but also concentrated at junctions, junctions too are likely candidates for elements of wolves' cognitive maps. They are frequently encountered and anointed with distinctive odors, and they are loci of decisions about direction of travel.

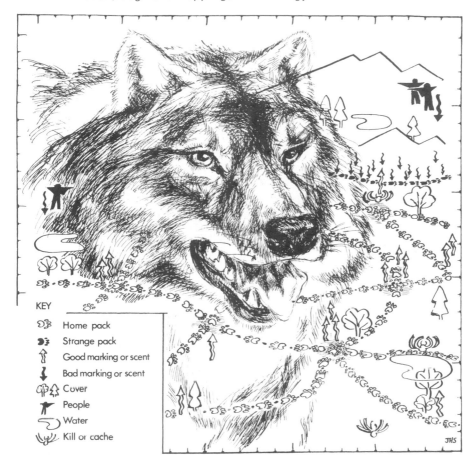

Figure 1. A wolf's cognitive map. (Drawing by John Slater.)

Kills are events of supreme importance in the lives of wolves. The locations of old kills are important not only because they can provide a small amount of nourishment when hunting has not been successful, but also because they mark places where hunting has been successful and may be successful again. Kills, surrounded as they are by several days' production of urine and feces, are salient sources of odor, readily detectable, even by the human nose, at some distance. They, too, are loci for decisions about where to go next. The combination of sensory distinctiveness and biological utility suggests that the locations of kills also may be encoded in the cognitive maps.

Wolves continue to place scent marks along routes with which they are familiar for reasons that have nothing to do with route choice. For example, scent marks are produced at higher rates along territorial edges, probably because of the proximity of foreign wolf-sign. Edges are also choice points, where packs decide to remain in their own territory or to trespass into an adjacent one.

The features of the environment that wolves mark with odors correspond to the features of the environment that they use, often insightfully, as routes, destinations, and choice points. These same features (locations, routes, and edges) occurred as elements and organizing principles in the representations of cognitive maps drawn by subjects whose tasks were to explore various natural or urban environments (Kaplan, 1973; Lynch, 1960; Peters, 1973).

During the summer of 1970 I helped Stephen and Rachel Kaplan conduct experiments on human cognitive maps of a natural area. Devlin (1973) presents a detailed description of these methods and results. About 30 subjects were led on a 20-minute walk through a variety of forest, meadow, and parklike environments. The hikers were then asked to draw maps showing as much as possible of what they remembered. Almost all such maps contained landmarks, regions, intersections of paths, and the forest edge, organized in a linear sequence by their placement along the line representing the path they followed.

About 25 different subjects explored the same area on their own, and each was then asked to draw a map. These maps included the same kinds of elements, but most preserved some spatial relationships among the major regions and landmarks. All but two of the rest were, like the maps from the previous project, based on the paths taken by the subjects. These paths, however, often formed a simple network rather than a line, and provided a basis for a more accurate portrayal of spatial relationships among the other elements than did the linear maps of the first subjects.

A third method of organizing the elements of a cognitive map was found in some of the maps of another area drawn by test subjects. In these maps, the first elements drawn were the boundaries of the area, and the other elements were placed in relation to these edges.

The same kinds of elements appeared in all the maps: paths, nodes (e.g., intersections), regions, landmarks, and edges. Similar results were obtained in urban areas by Lynch (1960) and De Jonge (1963). Our results further resembled Lynch's in the predominance of paths as organizing features.

The emergence of similar kinds of environmental elements (routes, locations, and regions) in human conceptions and lupine use of natural environments suggests that the ability to perceive and remember these elements may be cognitive adaptations to efficient use of a large range. The existence of the same kinds of elements in the movements and scent-marking of other species whose ways of life resemble those of wolves and hominids lends credence to this hypothesis. Routes, their junctions, locations of kills, and areas where prey can be killed are essential elements in the movements and scent-marking of all the other social carnivores: lions, hyenas, and hunting dogs (Kruuk, 1972; Lawick-Goodall & Lawick-Goodall, 1970; Schaller, 1972).

These elements may, of course, occur in the cognitive maps of nonhunting species. Because their food is less concentrated than that of carnivores, they do not travel long distances. Rather, they spend much of their time wandering in relatively small areas where the browse is good. Their cognitive maps, therefore, may not be as well developed as those of men, wolves, and, presumably, early hominids.

The extent to which olfactory sign facilitates wolves' recognition of routes and choice points within their territories is impossible to assess, but the regularity with which they are produced and investigated by traveling wolves suggests that they are somehow important. Early hominids, who probably did not use odors to create, in Kaplan's phrase, a "blunt sensory contrast" between landmark and surroundings, probably depended more on "refined sensitivity to the character of the surrounding area [1975, p. 109]" to recognize routes and locations. Travel in a large area thus probably made extra demands on their already well-developed powers of visual discrimination and memory. In the Pliocene, as today, one elephant trail probably looked very much like another, so the ability to recognize visual landmarks like rocks, trees, and even plant communities may well have been an important determinant of effective route choice.

STRATEGY IN WOLVES AND HOMINIDS

Peters and Mech (1975), in a discussion of hunting strategies of selected mammalian predators, listed the following techniques: persistence, encirclement with coordinated stalk, ambush, driving, blocking of escape routes before a rush, and running in relays. All these strategies have been reported by observers of hunting by wolves, and several were observed by Peters and Mech, who concluded that some or all of them may have provided early hominids with opportunities to exercise intellectual abilities.

However, none of these strategies seems to make demands on memory and planning as great as those of route choice in a large range. Such decision making involves not only knowledge of destinations, routes, and landmarks but also the weighing of probabilities and values, such as prey availability and time away from young.

This analysis of strategy in hunting suggests that the principal early sources of selection for intellectual abilities among hunting hominids lay not so much in wolflike strategies for the apprehension of prey as in wolflike knowledge of terrain. As Kaplan (1973) has suggested, the ability to form cognitive maps of terrain would be expected to transfer to other tasks, perhaps including tool use and toolmaking. An example from a field of modern human decision making that is generally regarded as extremely complex and extremely dependent on a large store of knowledge may help to illustrate the formal similarity between a cognitive map and a nonspatial plan of action. Several studies of complex decision making under uncertainty have emphasized the utility of treelike structures in representing relationships among decisions and their outcomes. An example of such a structure is shown in Figure 2, which illustrates the options available to a physician faced with a clinical decision involving a patient who may have high blood pressure and/or kidney damage. The physician may operate, use drugs, or seek more information. In such complex situations, some decisions are recursive, and the tree becomes a network; such a complexity is not easily illustrated, however, and any diagram understates the potential of the system represented. For example, in Figure 2, the right branch from node B (labeled *drugs*) should connect with node D, as the outcomes and subsequent decisions are the same. The analogy to a hunter who can follow a set of tracks, beat the bush, or explore further should be clear. His decisions to take a particular route or search a particular region have ramifications for this travel, which formally resembles the path chosen by the physician. Presumably the genetic systems of hunters who "thought ahead" in order to minimize their expenditure of energy while maximizing their probability of apprehending prey and returning home with it were favored by natural selection.

STRATEGY AND COGNITIVE MAPS IN HOMINID COMMUNICATION

Chimpanzees have been able to acquire a degree of fluency in communicating through American Sign Language, plastic tokens, and a modified computer terminal (Fouts, Mellgren, & Lemmon, 1972; Premack, 1971; Rumbaugh, 1977; Rumbaugh *et al.*, 1973). Their accomplishments have depended on rigorous training by humans but they show that, given the appropriate circumstances for a sufficient length of time, hominids with a chimpanzee's facility in manipulating symbols could have developed a rudimentary language. The major ecological difference between chimps and putative hominids like *Australopithecus* is that the former is a forager, whereas the latter was a hunter–gatherer. It is natural, therefore, to look to the ecology of the hunter–gatherer for clues about the problems and opportunities that might have produced circumstances appropriate for the development of language.

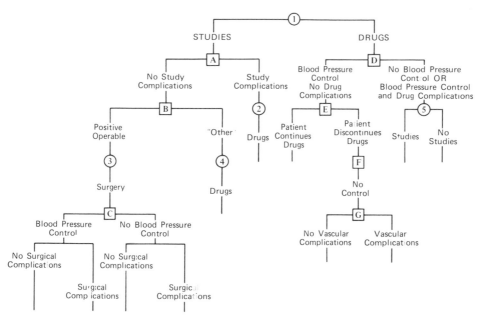

Figure 2. This decision tree describes possible actions a physician may take and their potential consequences in a patient thought to have either high blood pressure and/or kidney damage. The circles represent decision nodes, points at which the physician must choose an action; the squares represent points at which the outcome may go several ways but is not under the physician's control. This chart provides a model of a complex decision-making process and indicates a similarity between a geographic or spatial kind of cognitive mapping and a nonspatial plan of action. (Adapted from W. B. Schwartz, G. A. Gorry, A Essig & J. Kassirer, Decision analysis and clinical judgment *American Journal of Medicine*, 1973, 55, 459–472.)

Language has two aspects, the functional and the structural. Functionally, language refers to objects and events in the environment. Structurally, language has a grammar, that is, a system of rules for ordering symbols. Cognitive maps probably provided much of the content of early discourse, and perhaps they provided the structure as well.

Social hunting, with its dependence on coordinated strategy and movement, may have provided both the need and the opportunity to communicate about cognitive maps. If a group of hominids could learn to associate particular vocalizations (or gestures) with particular routes and locations they had visited together, they would have a means of communicating about important shared representations. ("Og is at the big tree.") It may not be accidental that wolves engage in a form of reference when they investigate odors that they have applied to their bodies by rubbing (Peters, 1973). In this way they convey the information that they have encountered a particular kind of object. If the wolves who investigate one of their fellows bearing such an odor have themselves visited the place where the odor originated, they would not only learn that he had encountered a smelly object, but might infer that he had visited a particular place. Hominids may have used sounds in similar ways.

Cognitive maps may have thus provided the structural as well as the functional foundations of language. Cognitive maps promote the ability to construct with ease sequences of representations of routes and locations. Presumably this ability transfers to sequences of other kinds of representations, involving objects other than places, and actions other than travel. The ability to combine representations of places quickly and easily may have reinforced (and been reinforced by) the ability to manipulate sequences of representations of actions directed toward the apprehension of prey. Once hominids had developed names (or other symbols) for places, individuals, and actions, cognitive maps and strategies would provide a basis for production and comprehension of sequences of these symbols. Shared network-like or hierarchical structures, when externalized by sequences of vocalizations or gestures, may thus have provided the structural foundations of language (Lamb, 1966).

In this way, cognitive maps may have been a major factor in the intellectual evolution of hominids. Forced by their predation on large animals to cover ranges much larger than those used by any other primates, they developed cognitive maps that were correspondingly more complex. Impelled by their lack of natural weapons to rely heavily on coordinated travel and strategy, they had a need to communicate that was greater than that of foragers. Lacking special odors and sensitive equipment to perceive them, they used sounds as names of places and of group members. Once this step had been taken, cognitive maps provided the structure necessary to form complex sequences of utterances. Names and plans for their combination then allowed the transmission of symbolic information not only from individual to individual, but also from generation to generation.

REFERENCES

Adler, H. E. Ontogeny and phylogeny of orientation. In L. R. Aronson, E. Toback, D. S. Lehrman, & J. S. Rosenblatt (Eds.), *Development and evolution of behavior.* New York: Freeman, 1970.

Bateson, G. *Steps to an ecology of mind.* New York: Ballantine, 1972.

De Jonge, D. Images of urban areas. *Journal of the American Institute of Planers,* 1963, *29,* 266–276.

Devlin, A. S. Some factors in enhancing knowledge of a natural area. In W. R. E. Preisser

(Ed.), *Environmental design research*. Stroudsburg, Pennsylvania: Dowden, Hutchinson, and Ross, 1973. Pp. 118–127.

DeVore, I., & Washburn, S. Baboon ecology and human evolution. In F. C. Howell & F. Bourliere (Eds.), African ecology and human evolution. *Viking Fund Publications in Anthropology*, 1963, *36*, 335–367.

Fouts, R. S., Mellgren, R. L., & Lemmon, W. B. American Sign Language in the chimpanzee: Chimpanzee to chimpanzee communication. Paper presented at the Ninth International Congress of Anthropology and Ethnology, Chicago, 1973.

Kaplan, S. Cognitive maps in perception and thought. In R. Downs & D. Stea (Eds.), *Image and environment*. Chicago: Aldine, 1973. Pp. 63–78.

Kaplan, S. Adaptation, structure, and knowledge. In R. G. Gollegde and G. T. Moore (Eds.), *Perspectives in environmental cognition*. Stroudsburg, Pennsylvania: Dowden, Hutchinson, and Ross, 1975.

Kinsey, A. C., Pomeroy, W. B., & Martin, C. E. *Sexual behavior in the human male*. Philadelphia: Saunders, 1948.

Kruuk, H. *The spotted hyena*. Chicago: Univ. of Chicago Press, 1972.

Lamb, S. *Outline of stratificational grammar*. Washington, D.C.: Georgetown School of Language, 1966.

Lawick-Goodall, H. van, & Lawick-Goodall, J. van. *Innocent killers*. New York: Ballantine, 1970.

Leyhausen, P. The communical organization of solitary mammals. *Symposium of the Zoological Society of London*, 1965, *14*, 249–263.

Lynch, K. *The image of the city*. Cambridge, Massachusetts: MIT Press, 1960.

Mech, D. *The wolf: The ecology and behavior of an endangered species*. Garden City, New York: Natural History Press, 1970.

Michael, R. P., Keverne, E. B., & Bonsall, R. W. Pheromones: Isolation of male sex attractants from a female primate. *Science*, 1971, *172*, 964–966.

Peters, R. P. Cognitive maps in wolves and men. In W. R. E. Preisser (Ed.), *Environmental design research*. Stroudsburg, Pennsylvania: Dowden, Hutchinson, and Ross, 1973.

Peters, R. P. *Wolf-sign: Scents and space in a wide-ranging predator*. Unpublished doctoral dissertation, University of Michigan, 1974.

Peters, R., & Mech, D. Behavioral and intellectual adaptations to the hunting of large animals in selected mammalian predators. In R. Tuttle (Ed.), *Socioecology and psychology of primates*. The Hague: Mouton, 1975. Pp. 279–300.

Premack, D. Language in chimpanzees. *Science*, 1971, *172*, 808–822.

Read, C. *The origin of man*. New York: Cambridge Univ. Press, 1923.

Rumbaugh, D. M. (Ed.). *Language learning by a chimpanzee: The Lana Project*. New York: Academic Press, 1977.

Rumbaugh, D. M., von Glasersfeld, E. C., Warner, H., Pisani, P., Gill, T. V., Brown, J. V., & Bell, C. L. A computer-controlled language training system for investigating the language skills of young apes. *Proceedings of the Ninth International Congress of Anthropology and Ethnology*, 1973.

Schaller, G. *The Serengeti lion: A study of predator–prey relations*. Chicago: Univ. of Chicago Press, 1972.

Schwartz, W. B., Gorry, G. A., Essig, A., & Kassirer, J. Decision analysis and clinical judgment. *American Journal of Medicine*, 1973, *55*, 459–472.

Tolman, E. Cognitive maps in rats and men. *Psychological Review*, 1948, *55*, 189–208.

6

Wolf Vocalization[1]

Fred H. Harrington and L. David Mech

Research into communication in the wolf has lagged far behind other wolf work, usually occurring only incidental to ecological investigations. However, during the mid 1960s a few studies—including ours involving radio-tagged wolves in northeastern Minnesota from 1972 through 1974 (Mech 1974; Mech & Frenzel, 1971)—have focused specifically on vocalization. As a result, at least general descriptions of the wolf's vocalizations, their usual contexts under natural conditions, and some understanding of their functions have finally emerged.

Many different sounds have been listed in the literature: barks, howls, growls, squeaks, whines, whimpers, songs, snarls, yelps, yips, yaps, and such combinations as growl-barks, bark-growls, and bark-howls. Unfortunately, in most accounts the name of the vocalization is the description as well; thus comparing sounds described by several authors becomes difficult. Theberge and Falls (1967) pared the list to six basic types; Joslin (1966) used four—howl, bark, whimper, and growl—and considered the others merely subclasses of the four.

THE GROWL

The growl is a deep, coarse sound, with energy spread between 250 and 1500 hertz (Hz) and emphasis around 800 Hz (Tembrock, 1963). (See

[1] This study was supported by the State University of New York at Stony Brook, the U.S. Fish and Wildlife Service, the U.S. Department of Agriculture North Central Forest Experiment Station, the Ober Charitable Foundation, the New York Zoological Society, and the World Wildlife Fund.

Figure 1.) Young (1944) described ''a deep snarl produced by exhaust of air through the wolf's partially opened mouth [p. 77].'' The snarl has been described in similar contexts as the growl (Fox, 1971), often in the same vocalization (i.e., growling snarls), and can probably be considered a form of the growl, since the growl may last only a few tenths of a second (as in a snarl) to several seconds. Growls are probably audible at less than 200 meters (Joslin, 1966).

Several students of captive wolves have described the role of the growl as a threat or warning (Fentress, 1967; Fox, 1971; Schenkel, 1947). ''Incipient attack,'' a preliminary to attack, consists of body shoves coupled with growling; milder forms of agonistic behavior, such as assertions of dominance or the defense of food or an object, are also often accompanied by growling.

During some forms of defensive behavior, such as ''protest snapping'' or the ''bite threat,'' growling may occur along with raising of the tail, stiffening of the legs, and bristling of the back hairs, according to these authors, but usually only in higher-ranking animals. Actual attacks sometimes follow if this behavior occurs with growling, leading Schenkel (1947) to conclude that growling serves as a threat, usually from high- to low-ranking animals. Very subordinate animals show few elements of threat behavior—that is, they keep their tails down or tucked under and their legs bent, and their manes do not bristle—and they do not growl. Growling also occurs when pups are play-fighting; again, the dominant animal (as defined by other criteria) growls more frequently and whines less frequently than the subordinate (Fox, 1971).

Between pups and adults, growling appears to be important when adults prevent or restrict a pup's behavior (Fox, 1971). For example, a female may discourage nursing by growling as the pups approach, and overexuberant play by pups with the adults is often halted with a growl. During danger, a growl or growl-bark (bark-growl) may send the pups quickly back into their den.

In field studies, records of growling are infrequent, probably because of its limited range. Joslin (1966) was often approached by one or more wolves when he howled near their summer homesites. Sometimes the animals growled repeatedly at him. In an attack of a wolf on a human, the wolf, considered rabid, was described as ''growling and gnashing his teeth'' through the attack (Rutter & Pimlott, 1968, p. 26).

Because growling is associated with agonistic behavior by high-ranking individuals, it appears to assert dominance, threat, or challenge. It seems to be a distance-increasing vocalization, either in reaffirming social distance (putting a subordinate in its place) or increasing actual distance between two animals (as in protection of an object, or in ''discipline'' or warning of the pups).

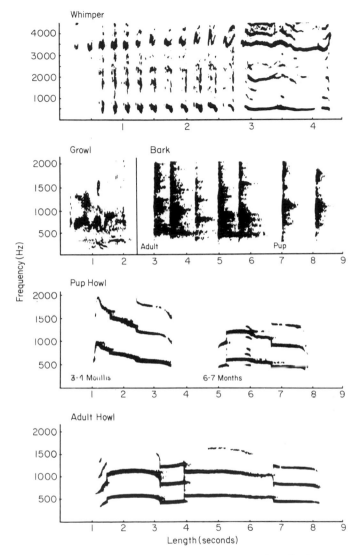

Figure 1. Wolf vocalizations. The sonagrams were prepared on the frequency ranges shown using a 150-Hz filter.

THE WHIMPER

Whimpering includes vocalizations variously classified as whines, whimpers, and squeaks. All are characterized by their high pitch and relatively pure tone. Early descriptions of whimpers were often very subjective, for example, ''a high, though soft, and plaintive sound similar to the whine of a puppy dog [Young 1944, p. 77],'' or ''a long, fervent string of mingled crying and wowing, hovering around one pitch [Crisler 1958, p. 150].''

The vocalizations in Figure 1 are characterized by energy between 400 and 800 Hz, but show an emphasis of energy at about 3500 Hz, giving them a high-pitched quality. (Frequencies given are the ''fundamentals,'' or lowest frequencies; most sounds also include components at multiples of the fundamental, called *overtones*, which contribute to the quality of the sound.) The duration of each burst of energy is from about .2 second to several seconds. The shorter sounds are probably what most workers have termed *whimpers*, whereas the longer, drawn-out ones are evidently the whines. The whimper sequence in Figure 1 is presented as it was recorded. The initial two vocalizations lack the low pitch of the whimpers, and to the human ear appear as squeaks. Field (1975) described the squeak as a high-pitched sound usually of less than 1 second. The average pitches of squeaks she recorded in a common context were 2500 Hz for an adult male, 2800 Hz for an adult female, and 3800 Hz for two pups, although pitch was variable, especially for the pups. Young (1944) stated that ''as the pitch of this whine varies so much, it seems ventriloquial [p. 77].''

Fentress, Field, and Parr found that squeaks from one animal varied in form as the context changed. For example, squeaks made during group howling were longer than those uttered when one animal approached another. Within the same context, however, squeaks from several animals (and even from another canid) on different occasions seemed strikingly similar in form. Thus frequency and duration of squeaks may be partly determined by their context, indicating that specific squeaks may have specific meanings.

Whimpers are usually fairly soft, and audible at no more than 200 meters (Joslin, 1966). Thus, like growls, their significance is probably most important within the pack in several contexts, which we shall discuss next.

Adult to Pup.   Schönberner (1965) described an adult male whining while pups were around him. Young (1944) reported that the alpha female used whines near the opening of the den, particularly if young pups were playing nearby. He felt that the whine indicated solicitude for the pups. Fox (1971) reported that ''undulating long whines'' brought the pups out of the den.

Pup to Adult.   When pups are cold, hungry, or in pain, they often whine, apparently to solicit care from adults (Fox, 1971). Pups whine and whimper as they approach adults, especially during greeting ceremonies that occur after an adult returns to a homesite (Peterson, 1974; Voigt, 1973). They have even whined when approaching humans who have howled near summer homesites. For example, Joslin's (1966) howling was answered by a nearby pup, which howled once, then rushed toward Joslin, whining twice. Upon encountering Joslin, the pup attempted to run around him, as if it thought the howling had come from behind him. Joslin concluded that the pup had mistaken his howling for an adult wolf's and it had rushed out to greet the animal.

Adult to Adult. Whines may occur during a broad range of nonaggressive behaviors, including the approaching and greeting of a high-ranking wolf by a subordinate, "paw raising" (often occurring before an approach during play solicitation), engaging in the mutual greeting ceremony, and the quick withdrawing and submission of a wolf during an agonistic encounter, when a sharp whine (yelp) may signal the end of the interaction (Fox, 1971; Schenkel, 1947).

Peterson (1974) once observed an adult returning to a recently abandoned homesite. As the animal entered the site, it whimpered.

Wolf to Human. Wolves socialized to humans often whine, whimper, or squeak, either when approaching these familiar humans or when being approached.

Whines Related to Sexual Behavior. Whine-type vocalizations also occur during courtship. Schenkel (1947) described a sexually receptive female as moving "with springing 'dance steps' while 'tenderly' whimpering or 'singing' [p. 93]." Peterson (1974) reported that whining between captive male–female pairs was especially common during the breeding season. A captive male wolf that one of us observed uttered many high-pitched, chirplike sounds shortly before copulation.

Whines during Chorus Howls. During group howls, individual animals—particularly young or subordinates—often whine or squeak (Field, 1975; Voigt, 1973), usually as they attempt to lick the face of another wolf (Zimen, personal communication).

It is evident that whimpers (whines, squeaks) are produced by all age classes and in many contexts. The underlying theme of all the behaviors is the friendly, nonaggressive attitude of the whimperer. These vocalizations occur when the vocalizer decreases its distance to another, either physically or perhaps socially.

The function of the whimper appears to be opposite to that of the growl, and this fundamental difference is underscored by facial expressions and body postures associated with each (Table 1). When the same animal alternates whining and growling during the same interaction, whining occurs with one set of behaviors, whereas growling accompanies another: "A wolf defending its prey from a conspecific or handler will successively

TABLE 1

Facial Expressions and Body Postures Associated with Whimper and Growl Vocalizations[a]

Component	Whimper	Growl
Head	Lowered	High
Neck	Extended horizontally	Arched
Ears	Flattened and turned down to sides	Erect and forward or flattened and turned back
Lips	Horizontal retraction	Horizontal contraction
Tail	Lowered or tucked	Raised
Direction of gaze	Away	Direct
Change in social distance	Decrease	Increase

[a] Based on Fox, 1970, 1971.

whine, wag its tail briefly and turn and lick or show licking intention, and then growl, elevate its ears and tail, and bare its teeth [Fox, 1970, p. 68]."

THE BARK

The bark is a short, explosive sound, usually no more than .1 second long (Figure 1). Tembrock (1963) gave the pitch of wolf barking as 320–904 Hz. The dominant pitch of the adult barks in Figure 1 is about 500 Hz. Often barks are uttered singly or in short, unpatterned sequences; however, in some contexts barks may occur in a series with a somewhat regular sequence, as discussed later in this section.

In captivity, barking has been described from animals attempting to solicit play from others; growl-barks have been emitted by adults sending pups into the nestbox (Fox, 1971). In captive pups, barking often occurs after return to a familiar place following transport to a new area (Fentress, 1967). Schenkel (1947) observed that during "defensive snapping" an animal may growl and bristle its hair and utter very sharp barking ("aggressive barking"). If the animal was very subordinate, however, Schenkel found that no barking accompanied the defensive snapping.

In the field, barking has been described in several contexts. Joslin (1966), Voigt (1973), and Peterson (1974) have all described chorus howls in which a sharp bark appeared to terminate the howling session. Joslin termed it the alarm bark. Other workers have heard barking after they approached close either to homesites containing pups (Joslin, 1966;

Murie, 1944; Theberge & Pimlott, 1969) or to kill sites (Mech, 1966). Joslin (1966) characterized the barking by its regular pattern: "two, occasionally one, sharp barks, followed by a more drawn-out bark which ended in a series of softer, lower pitched barks. . . . It was repeated 37 times over a period of 27 minutes. Following each series of barks there was a pause of approximately 50 seconds. Frequent growling occurred during this pause [p. 53]." Joslin termed this sound the threatening bark, and felt it was a warning or challenge to the individual barked at.

Barking is often interspersed with howling. In captivity, bark-howling usually occurs during some type of disturbance. Fox (1971) reported it from adults when a human entered a pen where there were pups; we observed it when a strange dog approached the pen of a captive wolf, as has Crisler (1958). Fentress could reliably elicit barks from a captive adult wolf if he howled within the wolf's pen after the animal was already howling. At a wolf compound in Nova Scotia, we have often heard the alpha female bark-howl when humans work in a nearby pen housing two wolf pups.

In the field, we heard bark-howling on two occasions after approaching and howling within 50 meters of a pack and a pair of pups that included radio-tagged members. Murie (1944) and Mech (1966) reported similar barking interspersed with howling during their close approaches to wolves.

In another case, two wolves apparently used bark-howling to meet each other. One animal barked and howled from one location, while the other howled, coming ever closer to the first. Finally, after the two met they moved off together, continuing to howl but no longer barking (Rutter & Pimlott, 1968).

The bark may serve to direct attention toward the vocalizing animal (Bekoff, 1974). It is perhaps significant that barking individuals are often visually conspicuous. Murie (1944) described the barking adults as moving about visibly in front of him. In one case, when he attempted to approach pups closely, two adults moved toward him barking, and when Murie left, one continued to escort him .8 kilometer ($\frac{1}{2}$ mile) away from the den. Haber (in Fox, 1971) had the impression that a barking wolf was attempting to decoy him away from the pups. In the example cited previously of wolves barking and howling to rejoin, the barking animal remained at its original location, drawing the other toward it. If barking does serve to draw attention to the vocalizer, then its physical characteristics, namely short bursts of energy covering a fairly broad spectrum (Figure 1), should make it relatively easy to localize (Konishi, 1973; Marler, 1955).

Joslin differentiated two types of bark, the alarm and the threat. Each was used in different contexts and also differed in quality. The alarm bark usually occurred alone, whereas the threatening bark occurred as a sequence. No rigorous analysis, however, has been conducted to determine whether there are predictable changes in the bark as context changes.

THE HOWL

The wolf vocalization that comes quickest to mind is the howl. Early descriptions reveal the impact that the howl has had on man: "The cry of the lobo is entirely unlike that made by any other living creature; it is a prolonged, deep, wailing howl, and perhaps the most dismal sound ever heard by human ear. I wish I could describe this howl, but the best comparison I can give would be to take a dozen railroad whistles, braid them together and then let one strand after another drop off, the last peal so frightfully piercing as to go through your heart and soul; you would feel as though your hair stood straight on end if it was ever so long [Young, 1944, pp. 76, 78]."

Objective studies, however, have been undertaken only since the mid-1960s. Using three adult male timber wolves raised in captivity (each studied separately), Theberge and Falls (1967) recorded some 700 howls. They also collected data on factors that might influence the howl and its rate of production. They found that the fundamental pitch, the lowest pitch in the howl, was also the loudest, so they based their analysis on the fundamental. The beginning (first .5 second), ending (last .5 second), and midsection (the howl between beginning and ending) were studied for structure (pitch rises, falls, and so on), and pitch. The number and relative strength of harmonically related overtones were also determined.

Theberge and Falls described the howl as a continuous sound, from .5 to 11.0 seconds long, with a fundamental pitch from 150 to 780 Hz, and up to 12 harmonically related overtones. We found similar results from free-ranging adult wolves in Minnesota. For 155 adult howls, the fundamental lay between 150 and 640 Hz, whereas duration ranged from .6 to 8.2 seconds. Pup howls, however, averaged much shorter and higher, ranging up to at least 1360 Hz (based on 371 howls).

Two of the wolves studied by Theberge and Falls rarely produced more than 3 or 4 harmonics, but the third usually produced up to 6 and several times up to 12 harmonics. Though we were rarely close enough to the wild wolves in Minnesota to record the much weaker high harmonics, on one occasion we recorded 6 harmonics in a pup howl from about 50 meters away.

Howling Sessions

Joslin (1966) found that when a single wild wolf replied to his howling, the bout of howling usually lasted for about 35 seconds, during which the wolf uttered several howls. Theberge and Falls (1967) made similar observations from their captive animals. In Minnesota, we found that individual reply bouts might last less than 10 seconds (only 1 howl produced) to many minutes. One adult howled 14 times during a 4-minute period and 8 times in $2\frac{1}{2}$ minutes, whereas another replied 9 times in 3 minutes and 24 times in $8\frac{1}{2}$ minutes. Another adult, howling spontaneously (i.e., not to any known stimulus), howled 30 times in $3\frac{1}{2}$ minutes.

When several wolves in a pack howl, the vocalization occurs as a chorus. Joslin (1966) described the chorus howl as started by one wolf. After that animal's first or second howl, the rest of the pack usually joins in. Each individual starts with a few long, low howls, and works up to a series of short, higher ones somewhat in chorus with the other pack members. Group howls recorded in Minnesota followed the same pattern, although if the wolves were separated from one another (out of visual contact), the howls normally remained similar throughout the response, and did not show the great variability of howls in typical group sessions. Joslin found the average duration of a group howl to be 85 seconds, whereas those in Minnesota averaged only 60 seconds (range: 30–123 seconds).

Once a pack howls, it is often not possible to stimulate a second reply during the next 15 to 20 minutes (Joslin, 1966; Pimlott, 1960; Voigt, 1973). However, we have found that sometimes packs will reply immediately after their last response, even two or three times in succession. On other nights, a pack may respond only once, and despite our attempts over the next hour or more, will not reply again. The factors influencing this varying responsiveness have not been elucidated, but we are continuing to study the problem.

Daily and Seasonal Trends in Howling

Spontaneous group howling sessions (those not stimulated by known external factors) follow both daily and seasonal trends in frequency. Studies with both captive (Zimen, 1972) and wild (Rutter & Pimlott, 1968) wolves have shown peaks in howling during evening and morning. Peterson (1974) observed wolves at a rendezvous site on Isle Royale for 363 hours in 1973, and heard howling 62 times, mostly between 1900 and 1000 hours. While monitoring two rendezvous sites in Minnesota for a total of

2050 hours, we found that most of the spontaneous howling occurred between 2000 and 1000 hours (Figure 2). However, although the majority of howling may occur between dusk and dawn, howling has been heard at all times of day (Joslin, 1967; Mech, 1970).

Spontaneous howling in captive wolves increases from fall through win-

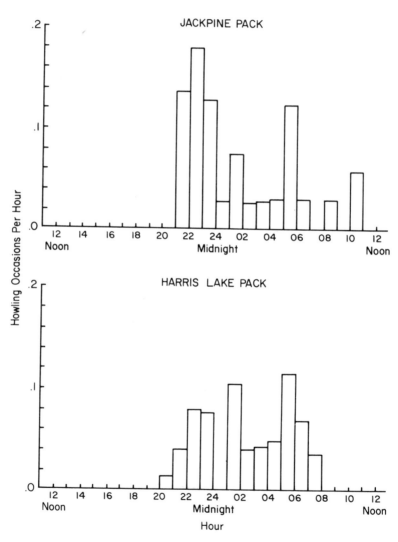

Figure 2. Time of day when spontaneous howling occurred around two rendezvous sites in Minnesota.

ter, usually peaking around the breeding season in midwinter (Fentress, Field, & Parr, in press; Klinghammer, 1975; Zimen, 1972). Stimulated or elicited howling from wild wolves shows similar seasonal trends (Figure 3). The increase in response rate in midwinter is correlated with the breeding season. When data from only sexually active animals (alpha males and alpha females) are considered, no responses occurred before the breeding season, but a high rate of response occurred during the height of the breeding season, around February 15–28 (Figure 4). Nonbreeding animals did not show such a sharp increase in response rate during the same period.

After the breeding season, howling rate drops, both in captivity (Zimen, 1972) and in the wild. From a very low rate in early summer, the response rate of wild wolves begins to rise in July and continues high throughout summer.

Joslin (1967) felt that the low response rate in early summer was in part due to the reluctance of adults to howl when the pups were small, since Schönberner (1965) and Rabb (in Joslin, 1967) had reported that adults would not howl until the pups were 9 and 6 weeks old, respectively. Joslin felt that this lack of vocalization would help protect the vulnerable pups from any possible predators. Voigt (1973), however, studying the same wolf packs as Joslin, several years later, found no increase in adult responsiveness after denning. Upon reexamining the earlier data, Voigt

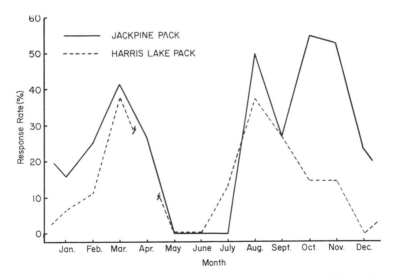

Figure 3. Response rates (to human howling) of wolves in two Minnesota packs throughout the year.

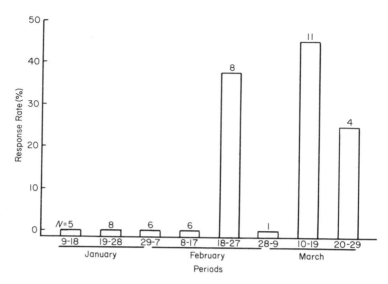

Figure 4. Response rate of alpha animals to human howling during 10-day periods in winter. Sample size is shown above each bar.

found that the increase in howling rate in July was not due to increased howling by the adults, but rather to the increased howling of pups. Data from Minnesota support this hypothesis, since contributions from pups greatly influenced response rate during July and August.

Development of Howling

Captive pups first begin to howl when 1 to 2 months old (Fentress, 1967; Mech, 1970; Schönberner, 1965). The early development of the howl is largely unknown. As Mech (1970) noted, "Eventually the whines became prolonged until they could be recognized as howls [p. 136]." In the field, pups in Ontario and Minnesota are usually first heard howling in early July, when 60–70 days old (Joslin, 1966; Voigt, 1973).

The maturation of the howl has not been studied in detail, but we have gathered some background data that will be useful to such an investigation. We analyzed each recorded howl for two variables—the average pitch of the fundamental (determined every .2 second) and the length of the howl. Though some overlap between adult and pup howls exists, pups usually tend to howl at a higher average pitch and for shorter durations (Figure 5).

Throughout their first 6 months of age, pups tend to howl increasingly deeper and longer (Table 2). We could not reliably separate the individual

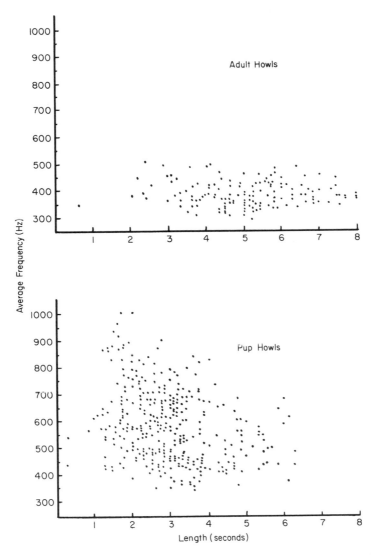

Figure 5. Characteristics of adult and pup howls recorded in Minnesota. Average frequency was determined by measuring the pitch of the fundamental every .2 second. Sample sizes: adults, 155; pups, 364.

TABLE 2

Characteristics of Individual Howls as a Function of Age[a,b]

	Pup howls			Yearling howls	Adult howls
	July/August	October	November		
Average frequency of fundamental (Hz)					
N	194	117	60	22	177
High extreme	1112	834	669	507	510
Median	585	540	528	372	372
Low extreme	398	346	360	346	301
Length of howl (seconds)					
N	191	118	59	22	154
High extreme	4.8	6.3	6.1	6.9	8.2
Median	2.5	3.5	3.1	3.9	5.1
Low extreme	.7	1.4	.3	1.7	.6

[a] Pups are born in late April or early May in general.

[b] For each howl, mean frequency of the fundamental and length of howl were determined. For each sample, the median howl and the high and low extremes are presented.

howls of older pups from individual adult howls, although generally there is an average difference. The foregoing conclusions were based on the pooled results from at least eight pups, two yearlings, and at least six adults. Thus individual differences and variation due to differential maturation are obscured.

A finer study was attempted using data from single animals. Unfortunately, data were available only for short periods from each animal. In all but one case, however, average pitch decreased as the pup grew older (Figure 6). The one exception will be examined later. Thus the limited data available at present indicate that pitch and length of howls change gradually during a pup's first half-year, and are little different from those of an adult by the time a pup is 1 year of age.

Individuality in the Howl

A fundamental question addressed by Theberge and Falls (1967) was individuality: (a) whether there are characteristics of the howl that can code individual differences and (b) whether wolves are able to detect these differences. They studied howl form, harmonic structure, and pitch preferences for each of three captive-raised adult male wolves from Algonquin Park, Ontario, Canada.

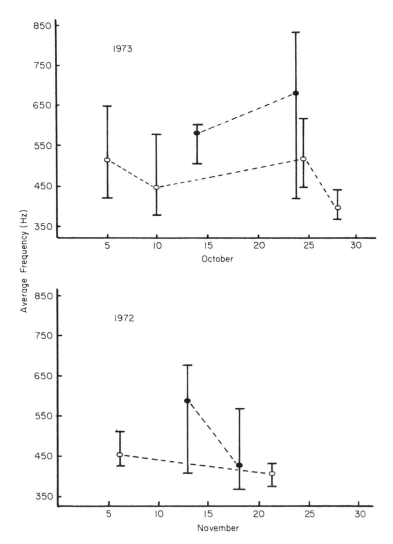

Figure 6. Average frequency (pitch) of howls by known, radio-collared pups. Circles represent the median howl in the sample, and bars enclose the range of the sample. 1972: solid circles represent male 2443; open circles, female 2445. 1973: solid circles, male 5065; open circles, male 5069.

Each individual had a tendency toward a certain type of beginning, ending, pitch range, and pitch change throughout the howl (Table 3). Furthermore, one had a different howl quality from the other two, as determined by harmonic structure. Thus, the howl of each wolf might

TABLE 3

Variation in Howling Characteristics of Individual Wolves[a]

Howling characteristic	Wolf A	Wolf B	Wolf C
Type of beginning (first .5 second)	Smooth rise in pitch	Break upward in pitch	Break upward in pitch
Type of midsection			
1. Sudden drops in pitch	No	0, 1, or 2	0, 1, or 2
2. Rise in pitch including highest note in howl	Yes	No	No
3. Rises in pitch not including highest note	No	No	0, 1, or 2
Type of ending (last .5 second)	Steady or no abrupt change	Slur rapidly down	Slur or steady
Average highest note	Middle A	C# high	D# high
Range of highest notes	14 semitones	5 semitones	8 semitones
Average lowest note	Middle C	—	F#
Range of lowest notes	19 semitones	—	9 semitones
Average length of howls	3.5 seconds	4.7 seconds	6.4 seconds
Range of lengths	8 seconds	6 seconds	11 seconds
Number of harmonics	Up to 12	Up to 5	Up to 5

[a] Adapted from Theberge and Falls, 1967, and reprinted from Mech, 1970.

possess a unique combination of such properties and therefore might be identifiable. Several field researchers have claimed the ability to distinguish specific wolves in a pack by their characteristic howls (Joslin, 1966; Rutter & Pimlott, 1968; Voigt, 1973).

With wild wolves in Minnesota, we also found individual differences between some animals. In one case, the howls of two adults in the same pack during a 2-week period in fall 1972 were compared. Two basic howl types were used by each wolf, a "flat" howl (coefficient of pitch variation less than 6%) and a "breaking" howl (coefficient of pitch variation greater than 10%). (See Figure 7.) For each type of howl, the howls by one adult were higher in average pitch than the howls of the other adult. In another case, two 6-month-old male littermates were howling from the same site but about 50 meters apart, so each howl could be assigned to a specific animal. In this case, there was virtually no overlap between the two pups when pitch and length were considered together (Figure 8). Thus, although gross frequency and time provide only rough measures of howl characteristics, the fact that individual wolves can sometimes be distinguished through these elements alone makes it more likely that, with the addition of other variables, individual recognition is possible.

To determine the ability of wolves to discriminate between similar sounds, Theberge and Falls (1967) presented live howls and tape

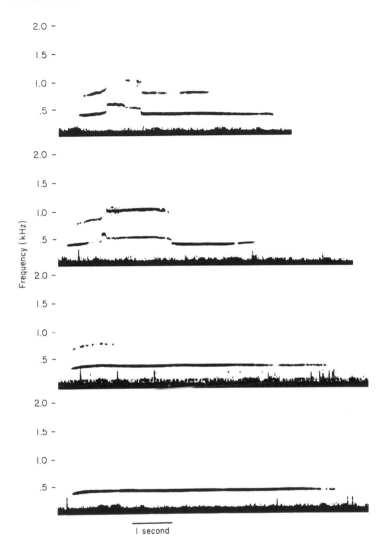

Figure 7. Sonagrams of howls recorded from an adult wolf in Minnesota. Upper two represent "breaking" howls (coefficient of variation > 10%); lower two represent "flat" howls (coefficient of variation < 6%).

recordings of the same howls to one of their captive wolves. The wolf responded to 90% ($N = 43$) of the live howls but to only 8% ($N = 25$) of the recorded howls. (In neither case was the sound source visible to the animal.) Upon spectrographic analysis, they found that the playback howl

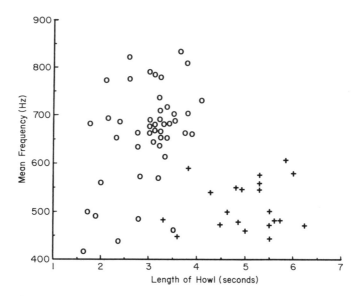

Figure 8. Comparison of howls of two Minnesota pups recorded the same night at the same site. Circles represent male 5065; crosses, male 5069. Sample sizes: for 5065, 46; for 5069, 21.

differed from the live howl in three qualities: *(a)* slight distortion of the fundamental, *(b)* reversal of relative strength of the first two harmonics, and *(c)* slightly less total volume. The results demonstrated the ability of a wolf to distinguish subtleties in sound, and strengthened the possibility that individual recognition could take place through howls.

Potential Information in Wolf Howls

It was mentioned earlier than the acoustic properties of squeak-type vocalizations appear to be correlated with context. The anecdotal literature has long assumed that each howl has a specific meaning (Crisler, 1958; Mowat, 1963; Young, 1944). Fentress (1967) claimed to be able to distinguish howls emitted when the animal was left alone from those emitted after feeding, and Rutter and Pimlott (1968) wrote of the "lonesome" howl. Further detailed documentation would be of considerable interest.

Theberge and Falls (1967) sought to determine whether certain qualities of a howl of three animals could be correlated with their type of movement (lying, pacing, walking slowly). They found that one howl characteristic was correlated with types of movement in two animals studied, whereas other traits were only correlated in one or the other animal (Table

4). In addition, they compared spontaneous versus elicited howls, and found eight significant relationships involving seven characters. Again, only one characteristic was significant in two animals.

Theberge and Falls concluded that communication could take place within a species at two levels, universal and individual; Universal communication "requires a symbolism that is the same throughout the species. . . . Communication on the individual level may occur between animals that have learned to recognize individual traits in animals with which they are associated [p. 335]." Thus, the relationships they found between behavior and howl form could convey specific information between wolves familiar with one another.

Universal relationships were not found as often as individual ones, but still may convey important information in the howl. One such relation was the finding that spontaneous howls tend to contain higher notes than elicited howls. In howls recorded from two pups in Minnesota, the howling was elicited on every night but one. On that night (October 24, 1973) the howls for each pup averaged 75–100 Hz higher than on a previous night (and a subsequent night in one case) (Figure 6). Thus in two Algonquin wolves (raised in captivity) and in two Minnesota wolves (both wild), spontaneous howls tended to be higher in pitch than elicited ones. We as yet have no hypothesis for this finding.

TABLE 4

Relationship between the Characteristics of the Howl and Several Behavioral Considerations[a,b]

Howl characteristic	Influence of behavior (lying, walking slowly, pacing)			Influence of elicited versus spontaneous howls	
	Wolf A	Wolf B	Wolf C	Wolf A	Wolf B
Beginning	—	NT	—	+	NT
Ending	—	—	—	+	—
Midsection					
Sudden drop	NT	+	+	—	+
Rise including highest note	—	+	NT	+	—
Rise not including highest note	NT	NT	—	—	NT
Highest note	—	—	—	+	+
Lowest note	—	NT	—	+	NT
Length	—	—	—	+	—

[a] From Theberge and Falls, 1967, pp. 335, 336.

[b] Symbols: +, relationship; —, no relationship; NT, no test.

Functions of Howling

A chorus howl is a highly contagious event within the pack (Crisler, 1958). Usually a high-ranking wolf, although not necessarily one of the alphas, starts howling, and the others quickly approach it and join in (Mech, 1970; Zimen, personal communication). The ceremony is accompanied by considerable tail wagging, whining, and face nuzzling by subordinates, and general friendliness. It has been suggested that these group howls may serve an important social function within the pack, much as greeting ceremonies and other rituals may (Figure 9).

Among other functions, howling serves as a means of assembly. Single captive wolves, separated from either mates or their pack, will readily howl (Young, 1944; Zimen, personal communication) and they often do so for hours at a time. Theberge and Falls (1967) reported that when a human observer, to whom the wolves were socialized, left the area of their single wolf, its spontaneous howling rate increased several orders of magnitude. Fentress (1967) found similar results with the addendum that if the wolf could hear the voices of the people who had left, howling increased even more. Thus separation from pack mates (or their substitutes) appears to be a strong stimulus for howling by single wolves.

In wild wolves, howling evidently serves to assemble separated pack members. Murie (1944) described eight occasions when wolves howled either after becoming separated from others while hunting or when entering abandoned rendezvous sites. On five of these occasions the animal was shortly reunited with other members of its group. Mech (1966) observed a similar episode after a pack had become separated during a hunt. One animal appeared to howl from a ridge (observation was from a plane), and a short time later other members of the pack had gathered on the ridge.

In Ontario, Rutter and Pimlott (1968) described three episodes in which howling apparently served to assemble members of a pack of wild wolves. They mentioned a particular type of howl they felt might have special importance in assembly: "This deep, single howl with a few barks is described perfectly in *The Wolves of North America* where it is said to be 'a call for assembling a group of wolves' [p. 136]." Young's description was "a loud, deep, guttural, though not harsh, howl [1944, p. 77]." Thus howling sometimes serves as a contact call between separated pack members, and the possibility of a special howl in this context is suspected.

In a similar situation, adult wolves on their way to rendezvous sites often howl, and are frequently answered by other adults and pups at the rendezvous sites (Peterson, 1974). Rutter and Pimlott (1968) described a similar situation involving four captive pups on an island. When a human

Figure 9. Wolf chorus. Among other functions, howling serves as a means of assembly. (Drawing by John Slater.)

howled from the lakeshore, the pups quickly answered, ran howling toward the person, then stopped and listened. Thus howls may serve to announce the imminent return of adults. Since these howls are often answered by animals at the rendezvous site, these replies may help guide the adults directly to the pups, especially if the pups are not at the usual location. In this respect, howling can be considered a form of assembly call.

Howling also seems to have significance between packs as well as within. Interpack howling sometimes continues for hours (Joslin, 1966; Voigt, 1973), promoting speculation that howling may function in territorial advertisement or maintenance (Joslin, 1967; Theberge & Falls, 1967). In Minnesota, three adjacent packs howled during the same session, each from within its own territory. After such sessions, packs often move apart, as described by Rutter and Pimlott (1968): "After a lot of howling with the Shanty Lake pack, both retreated, one east and one west [p. 141]." Such movements suggest that interpack howling occurs in an agonistic context, implying a warning or threat.

To function effectively in territorial maintenance, howling must be audible over the great distances often separating packs. Of the four basic types of vocalizations, howling has by far the greatest range. According to Joslin (1967), humans can sometimes hear wolf howls in Algonquin Park at a distance of 6.5 kilometers (4 miles), and Stephenson (in Henshaw & Stephenson, 1974) reported that humans can hear wolf howls at 16 kilometers (10 miles) on the North Slope of Alaska, where lack of vegetation optimizes sound transmission. In wooded areas of Minnesota, wolves apparently responded to human howling at distances as great as 10.8 kilometers (6.8 miles) in our study. Thus, on calm, quiet nights, a single howling session could advertise a pack's presence over an area of from 50 square miles (Joslin, 1967) to 140 square miles or more. Since most pack territories in Minnesota and in other forested areas range between 40 and 120 square miles (Mech, 1973; Pimlott, Shannon, & Kolenosky, 1969), howling can quickly cover a large area.

Therefore, two observations—that howling interactions sometimes occur between packs, and that after howling packs tend to avoid one another—provide circumstantial evidence that howling is used in some way in territorial maintenance. We are currently working on this hypothesis in detail and the results so far tend to support the hypothesis.

CONCLUDING REMARKS

The study of wolf vocalization is in its infancy. Detailed acoustic studies of each sound form, its variations, and possible correlations with behavioral and ecological contexts are lacking for most wolf vocalizations. For howling, some preliminary work has been completed, but we are still far from a full understanding of these rich and varied vocalizations.

ACKNOWLEDGMENTS

The cooperation of the Superior National Forest is gratefully acknowledged, as is the assistance of Jeff Renneberg, Glynn Riley, the late Robert Himes, and numerous student interns. Charles Walcott provided considerable logistical support and advice, and John C. Fentress made many helpful suggestions for the improvement of this chapter.

REFERENCES

Bekoff, M. Social play and play-soliciting by infant canids. *American Zoologist*, 1974, *14*, 323–340.

Crisler, L. *Arctic wild*. New York: Harper, 1958.

Fentress, J. Observations on the behavioral development of a hand-reared male timber wolf. *American Zoologist*, 1967, *7*, 339–351.

Fentress, J. C., Field, R., & Parr, H. Social dynamics and communication. In H. Markowitz & V. Stevens (Eds.), *Behavior of captive wild animals*, in press.

Field, R. A perspective on syntactics of wolf vocalizations. Paper presented at A. B. S. Wolf Symposium, North Carolina, 1975.

Fox, M. W. A comparative study of the development of facial expressions in canids; Wolf, coyote and foxes. *Behaviour*, 1970, *36*, 49–73.

Fox, M. W. *The behavior of wolves, dogs, and related canids*. London: Jonathan Cape, 1971.

Henshaw, R. E., & Stephenson, R. O. Homing in the gray wolf. *Journal of Mammalogy*, 1974, *55*(1), 234–237.

Joslin, P. W. B. *Summer activities of two timber wolf* (Canis lupis) *packs in Algonquin Park*. Unpublished master's thesis, University of Toronto, 1966.

Joslin, P. W. B. Movements and homesites of timber wolves in Algonquin Park. *American Zoologist*, 1967, *7*, 279–288.

Klinghammer, E. Analysis of fourteen months of daily howl records in a captive wolf pack. Paper presented at A. B. S. Wolf Symposium, North Carolina, 1975.

Konishi, M. Locatable and nonlocatable acoustic signals for barn owls. *American Naturalist*, 1973, *107*, 775–785.

Marler, P. Characteristics of some animal calls. *Nature*, 1955, *176*, 6–7.

Mech, L. D. *The wolves of Isle Royale*. National Park Service, Fauna Series 7. Washington, D.C.: U.S. Government Printing Office, 1966.

Mech, L. D. *The wolf: The ecology and behavior of an endangered species*. Garden City, New York: Natural History Press, 1970.

Mech, L. D. *Wolf numbers in the Superior National Forest of Minnesota*. U.S. Department of Agriculture Forest Service Research Paper NC-97. Saint Paul, Minnesota: North Central Forest Experiment Station, 1973.

Mech, L. D. Current techniques in the study of elusive wilderness carnivores. *Proceedings of the International Congress of Game Biologists*, 1974, *11*, 315–322.

Mech, L. D., & Frenzel, L. D. (Eds.). *Ecological studies of the timber wolf in northeastern Minnesota*. U.S. Department of Agriculture Forest Service Research Paper NC-52. Saint Paul, Minnesota: North Central Forest Experiment Station, 1971.

Mowat, F. *Never cry wolf*. New York: Dell, 1963.

Murie, A. *The wolves of Mount McKinley*. National Park Service, Fauna Series 5. Washington, D.C.: U.S. Government Printing Office, 1944.

Peterson, R. O. *Wolf ecology and prey relationships on Isle Royale*. Unpublished doctoral dissertation, Purdue University, 1974.

Pimlott, D. H. The use of tape-recorded wolf howls to locate timber wolves. Paper presented at Twenty-second Midwest Wildlife Congress, 1960.

Pimlott, D. H., Shannon, J. A., & Kolenosky, G. B. *The ecology of the timber wolf in Algonquin Provincial Park*. Ontario Department of Lands and Forests, 1969.

Rutter, R. J., & Pimlott, D. H. *The world of the wolf*. Philadelphia: Lippincott, 1968.

Schenkel, R. Ausdrucks-studien an Wölfen. *Behaviour*, 1947, *1*, 81–129.

Schönberner, D. Observations on the reproductive biology of the wolf. *Zeitschrift für Säugertierkunde*, 1965, *30*(3), 171–178.

Tembrock, G. Acoustic behavior of mammals. In R. Busnel (Ed.), *Acoustic behavior of animals*. London: Elsevier, 1963. Pp. 751–783.

Theberge, J. B., & Falls, J. B. Howling as a means of communication in timber wolves. *American Zoologist*, 1967, *7*, 331–338.

Theberge, J. B., & Pimlott, D. H. Observations of wolves at a rendezvous site in Algonquin Park. *Canadian Field-Naturalist*, 1969, *83*, 122–128.

Voigt, D. R. *Summer food habits and movements of wolves* (Canis lupis l.) *in central Ontario*. Unpublished Master's thesis, University of Guelph, 1973.

Young, S. P. *The wolves of North America*, Part I. New York: Dover, 1944.

Zimen, E. *Wölfe und Königspudel*. Munich: Piper, 1971.

7

Scent-Marking in Wolves[1]

Roger Peters and L. David Mech

Scent-marking in mammals—the application of an animal's odor to its environment—has long intrigued researchers from various disciplines. However, because of a dearth of detailed knowledge about the behavior of free-ranging mammals, most research has been restricted to captive animals. Lack of adequate tracking and measuring techniques and the intrinsic difficulties present in studying olfaction have also greatly hindered such investigations. Thus little is known about the ecological and sociological context and implications of scent-marking under natural conditions, and detailed descriptions of the frequency and distribution of scent marks in the wild are available for only a few species.

Mykytowycz (1974), working with captive and free-ranging European rabbits *(Oryctolagus cuniculus)*, and Thiessen (1973), whose subjects were captive Mongolian gerbils *(Meriones unguiculatus)*, have gained remarkable insight into the scent-marking systems of these two species. Through histological, neurological, endocrinological, and biochemical studies, they have learned that *(a)* the development and use of scent glands are related to sexual maturity and the presence of gonadal hormones, *(b)* dominant males tend to scent-mark most frequently, and *(c)* scent-marking is related to possession of territory. Other researchers have pos-

[1] Reprinted, with slight revisions, from *American Scientist,* 1975, *63,* pp. 628–637, with permission of the publisher. Several figures and tables containing supporting data on scent-marking were omitted from this revision; the interested reader is encouraged to consult the original version.

tulated that some of these findings are true for other species under natural conditions, and there seems to be general intuitive agreement on these three points (Ewer, 1973; Ralls, 1971). However, few actual data from field studies have been available to confirm or elaborate these concepts.

In canids, scent-marking is a well-known phenomenon, commonly observed in domestic dogs, and there has been much speculation about its functions. Investigators (Mech, 1970; Schenkel, 1947) who have studied the social behavior of wolves *(Canis lupus)* and its ecological context have assumed that scent-marking is associated with territory maintenance. The Canadian naturalist–author Farley Mowat (1963), in a widely read fictionalized account of his interaction with wolves, *Never Cry Wolf,* popularized the notion that wolves produce a line of scent marks around their territory that neighboring packs do not dare cross.

Despite the history of speculations, assumptions, and conjecture linking scent-marking in wolves to territory maintenance, it has only recently been possible to gather hard data on the subject. This possibility arose as a result of intensive radio-tracking studies of wolves in the Superior National Forest of northeastern Minnesota (Mech, 1972, 1973, 1974; Mech & Frenzel, 1971). From 1968 through 1973, 96 wolves were radio-tagged, and the interactions of 13 contiguous packs were studied. Radio-tagging wolf packs provides three critical advantages for gathering data on scent-marking: *(a)* each pack can be identified, *(b)* each pack can be located at any time so that it can be tracked in the snow, and *(c)* a history can be developed for each pack, including number of members, sex and age of at least some members, and size and location of the pack's territory.

Wolf packs in the Superior National Forest are basically territorial, and most stable territories range in size from 125 to 310 square kilometers. Territories seem to be stable and exclusive from year to year under normal conditions, but over several months they may overlap about 2 kilometers along the borders. Interpack contact, however, is rare or nonexistent along the overlap.

The basic members of each pack are a dominant male and female—the alpha pair. Generally the alpha male is the pack leader, and he takes the initiative in leading attacks on prey and intruders and directs the movements and activities of the pack (Mech, 1970). The remaining pack members are usually the offspring of the alpha pair, from several litters. If an alpha animal perishes, one of the mature offspring probably takes its place.

Within the pack a definite social hierarchy develops below the alpha pair. The youngest litter of pups is subordinate to the other pack members, and among littermates some are dominant over others. Thus a pack can be viewed as a group of related, interacting individuals with various social ranks that keep them compatible. However, as younger members mature, they may not accept their positions. If they are low ranking, they may not tolerate being subordinate to the other members. Or, if they are high ranking, they may not accept domination by the alpha animals. In either case, the resulting disruption of the social order may lead to the departure of the individual from the pack.

Whatever the behavioral reason underlying the process, some but not all young wolves are forced to leave the pack and become loners. They may wander far from the pack territory and become nomadic in an area as much as 20 times the size of a territory (Mech & Frenzel, 1971). If during their wanderings they cannot avoid resident packs, they may be chased and attacked (Mech, 1970). If the lone wolves can find a suitable vacant area and a member of the opposite sex, they may mate and start their own pack. They probably determine whether a territory is vacant or occupied and find a mate by reading scent marks as they travel.

THE SCENT-MARKING STUDY

Scent-marking was studied during the winters of 1971–1972, 1972–1973, and 1973–1974 to help determine the role it plays in the information flow that is integral to maintaining the organization of the wolf population. The basic method was to track identified wolf packs in the snow and record the spatial and temporal frequency and distribution of the scent marks. Tracks were located by aerial or ground radio-tracking of the wolves. Observations of captive wolves supplemented the three winters of field studies.

During 1972–1973 and 1973–1974, we concentrated on two adjacent packs near our field headquarters, because we could correlate aerial observations of behavior with ground investigations of their sign. After watching the wolves from the air, we could follow their tracks on the ground to the area where we had observed them, which provided a check on both the aerial observations and the interpretations of sign.

Data were recorded on a detailed sketch keyed to a topographic map. A "sample" of data was defined as a set of tracks examined along the entire length of a given stretch, and 100 samples, totaling 240 kilometers of ground tracking, were analyzed.

Wolves scent-mark in several ways that make the odors they apply to the environment especially apparent (Kleiman, 1966). We considered four kinds of scent-marking: *(a)* raised-leg urination (RLU), *(b)* squat urination (SQU), *(c)* defecation (scats), and *(d)* scratching. During all snow tracking, we recorded the frequency and distribution of these four types of sign. Because raised-leg urination involves the frequent delivery of small amounts of urine, it can be considered primarily a scent-marking behavior rather than elimination. In contrast, defecation and squat urination may be for both elimination and scent-marking. Thus, although we recorded all eliminations, we regarded RLUs as providing the most unambiguous information about scent-marking.

Our observations at both the Brookfield Zoo near Chicago and the St. Paul (Como) Zoo in St. Paul, Minnesota indicate that only mature, dominant wolves, primarily the alpha male or female, urinate with raised leg (Woolpy, 1968). For example, 22 of 27 RLUs observed in the Brookfield pack, which contained several mature wolves, were performed by the alpha animals, 20 of them by the alpha male. Furthermore, over 60% of the RLUs were associated with self-assertion, snarling, growling, or biting, whereas this was not true of SQUs. Twice in the field, we established by radio-location that the tracks we followed were those of pups. We found several SQUs, but no RLUs, in over 60 kilometers of tracks, which supported the captivity observations that pups do not usually produce RLUs.

We could not measure the exact amount of urine excreted in an RLU, but we could simulate the mark by squirting snow with 5 cubic centimeters of colored water. We recorded 584 RLUs in the 240 kilometers of ground tracking, an average of 1 RLU per 450 meters, with a range from none in 7 kilometers to 20 in 1 kilometer. All but 4 RLUs were directed at particular objects, such as blocks of snow, trees, shrubs, rocks, snowbanks—and even a plastic bag. The objects marked were always conspicuous, either because they protruded from the snow near the wolves' route or because they lay on or across the route. Once marked, of course, they also bore an olfactory stimulus.

In more than 70 instances we noted that wolves traveling on roads had left a nose-shaped indentation in the snow, suggesting that they had sniffed snowbanks at the height of a typical RLU. Under about 30% of the indentations we verified that an RLU was present by blowing away a layer of snow to reveal traces of urine, but in the rest of the cases we could not be sure. Many of the indentations were associated with a fresh RLU mark.

The characteristics of an RLU mark imply that its major function is the production of a prominent, long-lasting olfactory and visual signal. Depositing urine well above ground level, on a snowbank or tree, for example, facilitates dispersal of odor by wind, increases the evaporating surface as the urine trickles downward, and minimizes chances that the mark will be covered by snow or washed away by rain. In winter, when an RLU contrasts clearly against the snow, it is visible several meters away. Placing only a small quantity of urine on many different objects increases the total effectiveness of the amount.

A pattern of signs frequently observed in the field during the breeding season was a combination of an SQU and an RLU. One observation from the air and six in captivity suggest that this pattern results from an SQU by a female, investigated by a male, who then performs the RLU.

Like RLU and scratching, defecation is often associated with certain kinds of behavior under conditions that suggest it has significance beyond elimination. The ability of wolves to deposit scats on prominent objects and in particular places indicates a degree of autonomic and central control. Furthermore, defecation by wild and captive wolves often occurs in emotional contexts. It is often difficult to classify scats as marks or eliminations, because neither the defecation posture nor the product has a characteristic marking form, as is the case with the RLU. Scats can be classified as marks when they are deposited on prominent objects (e.g., snowbanks, stumps, shrubs, and empty beer cans), when they are found in large concentrations accumulated over several months, and when they are marked with scratching or urine by the same wolf that deposited the scat.

Scats in the immediate vicinity of kills, in the absence of RLUs and scratching, are probably primarily eliminative, but of course they have prominent visual and olfactory properties as well. Scats found where wolves crossed a road, or at trail junctions, may have been deposited as marks, but it is difficult to be sure since elimination is always a factor in all scats.

The distribution of scats around rendezvous sites also suggests that scats are sometimes left as marks. Rendezvous sites are places where growing pups are left while adults hunt during July, August, and September. Large scats are sometimes deposited at strategic points around the site, on trails leading into the central area, and especially at nearby junctions. Such deposits evidently are left by adults, and they often contain as many as six separate scats, indicating repeated or multiple visits to a location.

Wolf scats, whether deposited as marks or not, are powerful sources of odor, sometimes detectable by humans up to 10 meters away, even when the air is still and the temperature below $-20°C$. They may also bear the odor of secretions from the anal sacs, which empty on both sides of the anal opening and which may give the scats an individual identity. Therefore it is not surprising that wolves are at least as interested in scats as in urine and scratching. We noticed that wolves walking along roads frequently veered to sniff at scats, many of which lay beneath several centimeters of snow.

Scratching, which may release odors from the glands in the paws, may, like RLU marks, be both an olfactory and a visual signal. It is often preceded by RLU or by orientation to an olfactory stimulus, and like RLU it varies in intensity. When scratching at high intensity, a wolf paws the ground with alternate motions of the stiffened right and left forelegs, each combined with a similar movement of the rear leg on the opposite side. Although scratching is usually associated with elimination, a wolf generally takes several steps away from the urine or scat before scratching, and the material thrown behind is almost never directed toward the excreta. Furthermore, only high-ranking wolves scratch, another indication that this activity has significance beyond mere elimination.

DISTRIBUTION OF OLFACTORY SIGN

Our descriptions of the spatial and temporal distributions and contingencies of various forms of wolf sign are based on 240 kilometers of ground tracking and 40 kilometers of aerial tracking, primarily of three wolf packs. In the ground tracking, in snow fresh enough to ensure observation of all sign, we recorded 1006 possible marks: 584 RLUs, 193 scats, 170 SQUs, and 59 scratches. The RLUs, by far the most widely distributed strong-smelling and long-lasting form of wolf sign found, can be considered the most important form of indirect olfactory communication.

Wolves make all types of marks and eliminations throughout the year, but the relative frequencies of marking during various seasons have been unknown. Because of the lack of lasting snow in our study area during the warmer months, we could obtain data by snow tracking only from December through March. Nevertheless, during this period, which includes the breeding season in late February (Mech, 1970), we discovered some interesting differences. The average rate of RLUs increased throughout the winter, from about 2.5 per kilometer in December and January to a peak of 3.5 per kilometer in late February, and then dropped to about 1 per kilometer in March. Although we do not know the rate for spring and summer, sometime before December there is a strong increase in the RLU rate. The SQU rate does not increase until late January. Like the RLU rate, it peaks in late February and drops to its base level in late March, which strongly suggests that both SQUs and RLUs are related to breeding. In contrast, the rates of defecation and scratching remained constant from December through March. All figures agree with those we have obtained from observing wolves in captivity.

The distribution of tracks reflects the major occupation of the wolf—travel in search of prey. The main features of wolf movements are an extensive, complex network of regularly used travel routes, including frozen waterways, roads, and trails, and concentrations of prey often more than 10 kilometers apart. The distributions of wolf sign in different types of environment and in the centers and at the edges of territories are important in trying to interpret the significance of these signs.

A high proportion of all four kinds of sign left along roads and trails was found at junctions, as also noted by Seton (1909) and Mech and Frenzel (1971). Both tracks and aerial observations show that wolves rarely loiter at junctions any longer than the few seconds necessary to sniff a bush or two or to leave an RLU or other sign. Yet on roads and trails about 40% of the RLUs and scratches and 50% of the SQUs and scats were found at junctions.

The type of travel route also influenced the proportion of different kinds of wolf sign found. The average frequency of the various signs was recorded *(a)* for regular travel routes such as roads and trails, *(b)* for cross-country or "bush" tracks, and *(c)* for routes on frozen lakes and waterways. We divided 104 samples of tracks into 140 segments, each composed of a continuous set of tracks in one of the three environments. Comparison of the RLU rate for 73 segments on roads and trails (3.4 RLUs per kilometer) with the rate for 55 segments in the bush (1.7 RLUs per kilometer) revealed a significantly greater tendency toward RLUs on roads and trails than in the bush ($t = 2.3$, $df = 126$, $p < .02$). Although speed varied in the two environments, the gait for almost all samples was "walking."

Of 28 segments of tracks that went through both environments, the number of RLUs per kilometer was greater on the road and trail parts than in the bush parts in 18 of them, and in only 8 cases was the reverse true; in 2 cases they were equal. A sign test on the 26 samples showed that the wolves increased their rate of RLU when traveling along an established route and decreased it when cutting cross-country through the bush ($z = 1.98$, $N = 26$, $p < .05$).

The mean number of RLUs per kilometer is much lower on frozen waterways (.4 per kilometer) than in the other two environments. This low rate is easy to understand: Almost all waterways on which tracks were followed were iced-over lakes, where there are few vertical objects to serve as targets. In fact, all 7 RLUs recorded were on rocks or weeds protruding above the ice, or on the shoreline.

This effect of target availability makes the difference between rates of RLU on regular travel routes versus bush routes even more striking, for targets are far more available in bush than on roads. (The rate of RLU was high even on unplowed roads, where there were no banks to invite marking.) Clearly, availability of targets does not explain the greater rate of marking along roads and trails. When there are many potential targets, none are salient.

Probably the most important areal difference found was that between the rate of scent-marking in the centers of wolf territories and at the edges. Each tracking sample was classified as to whether it lay in a narrow strip about 1 kilometer wide along the edge of a territory or whether it fell in the center. A few samples that had segments in both were deleted from the analysis. The samples along the edges averaged 2.67 RLUs per kilometer, whereas the samples in the centers averaged 1.27. The average length of track sample (2.4 kilometers) was the same for both edges and centers, but there were 6.5 RLUs per sample at the edges and 3.1 in the centers, a highly significant difference ($z = 6.3, N = 376, p < .001$).

The number of wolves making the tracks we followed was, predictably, a major determinant of the rates of sign production, although it did not correlate similarly for each of the four kinds of sign. There was a linear relationship between the average numbers of scats, SQUs, and scratches per kilometer and the number of wolves tracked.

The RLU rate, however, did not vary as much with the number of wolves as did the other rates. Spearman r's (a measure of degree of correlation) for the relationship between rates of production of scats, SQUs, and scratches and the number of wolves are 1.0, .8, and 1.0, respectively, whereas the r for the RLU rate and number of wolves is only .2. The lack of correlation between rate of production of RLUs and number of wolves should not be surprising; because only one alpha pair dominates each pack, regardless of pack size, the number of wolves that urinate with raised leg generally does not increase with pack size.

STIMULI FOR SCENT-MARKING

The conventional view of canid scent-marking is based on the notion of von Uexküll and Sarris (1931) that stimuli from unfamiliar conspecifics are the primary "releasers" for scent-marking, especially for RLUs. Our data indicate that although scent marks of alien wolves do evoke high rates of marking, they are not the usual releasers. On at least 10 occasions the members of one of our radio-monitored packs remarked the same 2.4-kilometer stretch of road, including the same junctions, even though our daily inspections indicated that no other wolves had marked it in the interim. Several times this pack remarked the road while the odor of the RLUs from the previous visit was still detectable by a human. Evidently neither complete fading of the odor of an RLU nor its masking by an unfamiliar wolf is necessary to stimulate re-marking.

In fact, the fresher an RLU is, the more likely it is to elicit further RLUs. The major determinant of the distribution of RLUs is not sign from unfamiliar wolves but the *repeated use of travel routes by the resident pack*. Regular use of scent-marked routes produces positive feedback, which increases the probability of re-marking, with the result that sign of the resident pack is especially dense in areas visited frequently.

The data also allow us to estimate how long an RLU continues to stimulate more RLUs under winter conditions. The base rate is reached after the marks are approximately a week old. The rate of response continues to decline as the scent marks age, but we have recorded marking in response to familiar sign as much as 23 days old, although the rate was low, indicating that the stimulus value of familiar scent marks probably drops to zero soon after 23 days of age. Of course weather conditions could cause considerable variation in these figures.

Responses to Neighboring Packs

Our findings indicate that each pack's territory is well marked throughout with its own scent and that the marks are renewed through positive feedback. But what is the response of a resident pack to the marks of a neighboring pack? Data on this subject are difficult to gather because human beings cannot identify by odor the marks of different packs; however, we have no doubt that wolves can detect such differences. In four instances we were able to track a pack of wolves when it encountered fresh sign of a neighboring pack. Twice we found that there was an abrupt and immediate increase in the rate of scent-marking by the pack when it encountered the alien sign—from no scent marks in over 2 kilometers to approximately 10 scent marks, including at least six RLUs, in the first kilometer after the encounter. In the other two instances, the rate of scent-marking was the highest we have ever recorded but it was impossible to determine a base rate of marking just before the encounter.

As a result of such high rates of marking in response to sign of neighboring packs, a concentration of marks accumulates along the borders of each territory. Some of the marks are those of one pack, and some are those of its neighbors, whereas many marks are made by both groups alternately superimposing their own scents. In effect, each territory is an olfactory "bowl" with the edges composed of high rates of the resident pack's marking interspersed with high rates of the neighboring pack's marks (Figure 1).

FUNCTIONS OF SCENT-MARKING

The picture that emerges from the description of scent-marking in free-ranging wolves is that each pack of wolves travels about its 125 to 310 square kilometer territory irregularly but reaches most parts of it at least every 3 weeks and probably more frequently. The wolves travel mostly on game paths, old logging trails, dirt roads, and other established routes with which each territory is interlaced and, at least in winter, encounter (and leave) a sign every 240 meters on the average, including an RLU each 450 meters.

At their usual rate of travel, 8 kilometers per hour (Mech, 1970), wolves encounter and produce an olfactory sign about every 2 minutes, including an RLU every 3 minutes. Even if they strike out cross-country to pursue

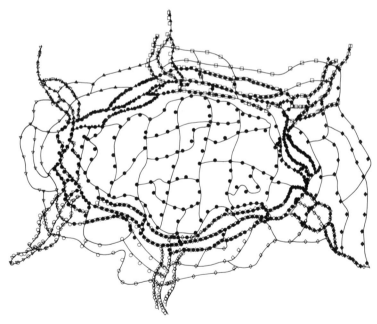

Figure 1. In this model of the distribution of RLU scent marks, RLUs are shown throughout the territory of one wolf pack (dark circles) and in the areas where six neighboring packs (indicated by all other symbols) border this territory. Travel routes are simulated, but mean density and territory size are to scale, on a range approximately 20 kilometers wide. Note the bowl-shaped effect caused by heavier scent-marking by both the resident pack and its neighbors at the edges of the territory.

prey or to take a shortcut, they leave sign frequently, although the RLU rate is depressed. This means that the entire territory is studded with olfactory "hotspots," and wherever a wolf is it can tell whether or not it is in its own territory. The concentration of scent marks around junctions further ensures that any traveling wolf will detect a mark in a short time no matter what route it uses. Each pack can also quickly detect when it reaches another pack's territory, and the nomadic lone wolves probably know just where they are in relation to the various packs' territories—on the border of two packs, in the center of a territory, and so on.

Because wolves seem to respond differently to scent marks of different ages, it appears that they can detect the time elapsed since they were last in any particular area. And perhaps the accumulation of a certain density of marks triggers travel to another part of the territory. At the border of the territory they probably can also tell approximately when their neighbors last passed through.

No doubt squat urination and defecation also carry such information. However, since they are basically eliminative in function and show few significant seasonal or spatial differences in distribution, and since all pack members, including pups, perform them, their scent-marking significance may apply primarily within the pack.

We have often seen pups and other pack members temporarily separated from the main pack. During summer, subordinates may spend days away from the den when the alpha pair attends the pups. Various pack members often hunt separately, especially during this period. By reading the urinations and defecations of their associates, they may be able to determine whether an area has been hunted recently, if an associate is nearby, or who is traveling with whom. This would ensure that efficient use is made of all parts of a pack's territory. Scratching, although usually done by alpha animals, particularly the male (as determined by our observations of animals in captivity), shows no seasonal or distributional variation, so its primary function might also be intrapack, perhaps as assertion of the alpha animal's continuing status.

It is raised-leg urination that is probably most effective in maintaining the pack's territory. The well-established relationship between RLUs and dominance, breeding, and territorial defense is circumstantial evidence that RLU scent-marking is intimately involved in territory maintenance. What is necessary for proof is evidence that one pack's RLUs cause aversion on the part of neighboring packs. Direct evidence for this is most difficult to obtain under field conditions, Farley Mowat's *Never Cry Wolf* (1963) notwithstanding. What observations we do have, however, are highly suggestive.

In one instance we found tracks of a pack of seven to nine wolves along the northwest edge of their territory. Their tracks proceeded southwest onto a frozen lake and along the northwest shore for about 2 kilometers. The wolves remained several meters from land but made 13 approaches to the northwest shore of the lake; each time they turned back before reaching land. The only type of aversive agent perceivable this far out on the lake that could have been distributed for such a long stretch along the shore would have been the scent marks of another pack.

In a second instance, eight wolves approached the east edge of their territory and crossed the tracks made by a neighboring pack of five wolves 2 weeks earlier. They scent-marked at the junction of the trails, some members followed the tracks for a short distance, and others continued eastward. Much scent-marking ensued, and after going about 1 kilometer the pack turned around and headed back into its own territory. If the aversive agent was not the odor of the neighboring pack, the most likely alternative explanation other than coincidence would be the lack of visual or olfactory familiarity with the area. This does not seem plausible, however, because each winter when the large lakes (some several kilometers across) in the study area freeze, and wolves venture far out onto the ice, even though this must also be relatively unfamiliar ground.

A third observation involved a river boundary between two packs. Soon after the river froze we found tracks of seven members of the pack north of the river crossing to the south shore and then returning. The next day, the south pack approached to within 40 meters of the river; one wolf approached the south shore five times and made RLUs and scratches at the northern extreme of each excursion. Where the south pack encountered the tracks of the north pack there was a network of wolf trails, covering 300 × 600 meters, in which there were 30 RLUs, 10 scratches, 2 SQUs, and 1 scate. The south pack went no further, but turned and headed back south.

In another, rather telling, instance involving these two packs, tracks showed that the north pack had chased a deer across the frozen river and wounded it severely. When the deer proceeded farther into the south pack's territory, the north pack did not persist after it as wolves usually do (Figure 2). Instead, the members scent-marked heavily in the area and then returned to their territory. A day later, the south pack located the deer and consumed it.

Obviously more such observations are necessary, and attempts must be made to watch the animals actually responding to the marks of their neighbors. Nevertheless, the present evidence is sufficient to allow us to formulate a hypothesis concerning the manner in which scent-marking (especially RLU), as the main information medium, helps hold together the spatial organization of the wolf population.

Figure 2. Wolves quitting the chase. (Drawing by John Slater.)

We do not regard scent-marking as an isolated system functioning independently of other behavioral traits and mental processes. Wolves appear to have well-organized memories for routes, points, junctions, and their juxtaposition. (See the chapter by Peters on cognitive mapping.) These cognitive maps, with which wolves travel their territories, probably also help them recognize territorial edges. In the four instances of avoidance just mentioned, cognitive maps, as well as unfamiliar sign, were probably involved. It is difficult to separate the effects of sign and terrain, since scent marks are found along all the major routes and at all important points in the territory.

Aversion to unfamiliar territory therefore may be involved in aversion to unfamiliar marks, and responses to foreign scent marks may depend on whether the wolves that encounter the marks are in their own area. Aversion to foreign odors probably is not innate; captive wolves we have observed did not avoid the odors of unfamiliar conspecifics. Nor is the response to foreign scent marks stereotyped, as the four examples cited demonstrate. The aversion to unfamiliar sign and territory may be acquired through rare agonistic encounters between packs or may be learned by exposure to emotional responses of adults who have been involved in such encounters.

At present the wolf population has reached the saturation point in our study area; there is no land unoccupied by wolves (Mech, 1973). In such a situation, we postulate that frequent scent-marking and aversion to strange marks hold each pack in its territory and that a system of positive feedback stimulus ordinarily keeps each territory adequately marked.

Several questions can be asked. What would happen if, because of unusual environmental conditions such as the reduction of prey in part of the territory, a pack neglected marking all its territory every 3 or 4 weeks? Or what would happen if a pack were exterminated? Presumably, the scent marks would eventually lose their stimulus value. Would this mean that the returning pack, or some other pack, would never again scent-mark the territory because of the lack of scent-mark stimuli? How are new territories set up?

There must be some sort of "setting" or "resetting" of the feedback system. Presumably, where there are no scent marks or when the stimulus value of scent marks reaches zero, the mere absence of marks must be a stimulus for wolves in the appropriate physiological and behavioral condition to mark. These wolves could be resident pack members returning belatedly to an unused part of their territory or perhaps a neighboring pack extending its own territory. They may also be a newly formed pair of loners that perceived the olfactory (territorial) vacuum.

Loners can easily locate one another by their scent marks. Since the SQUs of an alpha female are probably usually accompanied by the RLUs of the alpha male during the breeding season, each lone animal would be able to determine that the other was unmated. Just as within an established pack, the newly formed pair could then carry on a courtship—in which frequent scent-marking probably plays an essential role—could mate, bear young, and begin their own pack. Such a system would tend to ensure that all available habitat is occupied and that, if any territory were too large to be patrolled frequently enough, "surplus" animals would detect and colonize it. The ultimate result would be full use of available space and resources by a population that would in turn be regulated by the size of the colonizable area.

There are many gaps yet to be filled in our knowledge of the wolf scent-marking system. However, this basic description and hypothesis provide a good first approximation of the internal workings of a complex social organization, much like those proposed for other species (Mykytowycz, 1974; Thiessen, 1973), and also an excellent framework within which to pursue other detailed studies.

REFERENCES

Ewer, R. F. *The carnivores.* Ithaca, New York: Cornell Univ. Press, 1973.

Kleiman, D. Scent marking in the Canidae. In P. A. Jewell & C. Loizos (Eds.), Play, exploration and territory in mammals. *Symposia of the Zoological Society of London,* 1966, *18*, pp. 167–177.

Mech, L. D. *The wolf: The ecology and behavior of an endangered species.* Garden City, New York: Natural History Press, 1970.

Mech, L. D. Spacing and possible mechanisms of population regulation in wolves. *American Zoologist,* 1972, *12*(4), 9.

Mech, L. D. *Wolf numbers in the Superior National Forest of Minnesota.* U.S. Department of Agriculture Forest Service Research Paper NC-97. Saint Paul, Minnesota: North Central Forest Experiment Station, 1973.

Mech, L. D. Current techniques in the study of elusive wilderness carnivores. *Proceedings of the Eleventh International Congress of Game Biologists,* 1974, *11*, Pp. 315–322.

Mech, L. D., & Frenzel, L. D. (Eds.). *Ecological studies of the timber wolf in northeastern Minnesota.* U.S. Department of Agriculture Forest Service Research Paper NC-52. Saint Paul, Minnesota: North Central Forest Experiment Station, 1971.

Mowat, F. *Never cry wolf.* New York: Dell, 1963.

Mykytowycz, R. Odor in the spacing behavior of mammals. In M. C. Birch (Ed.), *Pheromones.* New York: American Elsevier, 1974.

Ralls, K. Mammalian scent-marking. *Science,* 1971, *171*, 443–449.

Schenkel, R. Expression studies of wolves. *Behaviour,* 1947, *1*, 81–129.

Seton, E. T. *Life histories of northern animals* (Vol. 2). New York: Scribner, 1909, Pp. 677–1267.

Thiessen, D. D. Footholds for survival. *American Scientist,* 1973, *61*, 346–351.

von Uexküll, J., & Sarris, E. G. Das Duftfeld des Hundes. *Z. Hundeforschung,* 1931, *1*, 55–68.

Woolpy, J. H. The social organization of wolves. *Natural History,* 1968, *77*(5), 46–55.

Part III

PALEOBIOLOGY

Ever since the period of Darwin, evolutionary theorists have argued the relative importance of two different kinds of evolutionary processes: *cladogenesis,* or branching evolution, and *anagenesis,* or progressive evolution. Though the terminology and focus have changed, the principal issue is the same: Is slow, progressive alteration of organisms responsible for most of the diversity in the animal world? Or is this vast diversity due to relatively few rapid, revolutionary changes in behavior and anatomy? We shall discuss both these concepts further in order to see how this issue shapes our understanding of the evolution of wolves and humans.

Cladogenesis, or branching evolution, is itself a complex concept that includes several kinds of events. In its simplest form it represents *geographic speciation,* the process whereby a species is physically subdivided—perhaps a population is cut in two by a river or by a break in its preferred habitat, as when a savannah develops in a forest and cuts off two branches of a tree-dwelling species. As a result of lack of genetic contact and divergent environmental pressures, the species divides into two.

A more complex and theoretically exciting type of cladogenesis is *adaptive radiation,* the proliferation of species that can occur when a population enters a new life zone. Adaptive radiation basically means one thing: an opportunity seized. An animal population may at first be only minimally adapted to a new life zone but when that zone opens up the population nonetheless accepts the challenge and makes the most of its minimal adaptation, improving upon it rapidly. A few examples will show the importance of behavior in this process: the proliferation of reptiles that occurred when that group colonized land, the radiation of birds when the "air zone" was entered, and the diversification of mammals in the Cenozoic.

Why does adaptive radiation occur? Many forces may be postulated, each operating under specific conditions. A new adaptive zone may open up because of a new behavioral innovation (flying in the case of birds, reproduction on land in the case of reptiles). Sometimes external conditions are responsible. For instance, when the giant reptiles died out at the end of the Mesozoic, those formerly rare animals, the mammals, suddenly were confronted with empty life zones that they now had the chance to fill. And fill them they did.

In contrast to cladogenesis, anagenesis refers to a gradual change of form within a lineage. When we talk about natural selection we usually refer to a process of this sort; mathematical formulations of selection and the branch of biology known as population genetics are generally concerned with selection within one biological lineage, hence with anagenesis.

Cladogenesis and anagenesis are terms of traditional usage but the theoretical argument now raging among evolutionists has generated its own terms. Eldredge and Gould (1972) in a chapter in the book *Models in Paleobiology* have used the term *phyletic gradualism* to refer basically to the process of anagenesis—slow, progressive change within a lineage over a long period of time. They use this term to cover not only the process but the opinion that gradual change is the most significant kind of evolutionary change, in the long run.

In opposition to phyletic gradualism, which they do not consider of major importance, Eldredge and Gould propose that evolutionary history should be described as *punctuated equilibrium*. By this they mean that evolution consists of long periods of essential equilibrium, involving little change, that are interrupted, or punctuated, by episodes of rapid species proliferation. They contend that new species arise on the periphery of the range of well-established species and that though most of the new peripheral populations are inadaptive and are selected out, some survive and prove immensely successful. Another term, *rectangular evolution,* refers to the same general process. This concept is discussed by Steven M. Stanley (1975).

One of the great commonplace untruths about science is that all one needs to do is go out and discover facts and let the facts speak for themselves. Nothing could be further from the truth of how scientists actually function and how scientific discoveries are made. Not one scrap of data on a subject such as the evolution of the hominids or the canids could be collected—let alone analyzed—without prior formulation of a theory for those data to test, or a framework within which to analyze the data. For years, the model guiding collection of data on most mammalian groups (including hominids and canids) has been the model of gradual, progressive evolution. The data gathered on the evolution of the early hominids, however, are beginning to upset the hypothesis of phyletic gradualism. Hence, in this book we shall take the position that we should try another interpretation—the model of cladogenesis, including adaptive radiation.

We must assume that major events of cladogenesis are guided by environmental changes as well as by the evolution of new behavioral and anatomical structures in animal populations. In the case of the hominids, the increase in savannah zones in Africa perhaps 5 million years ago offered an opportunity to a bipedal hominid and prompted the radiation of several species, and perhaps several genera, of hominids. Of these forms all but one have become extinct.

In the case of the canids, radiation probably was more recent and at least five major groups still survive: the wolf, the coyote, the jackal, the dingo, and the domestic dog. It is for this reason that the living diversity of the genus *Canis* may prove useful to us in reconstructing species interaction among extinct, diverse hominids. The next two chapters analyze the paleobiology of the North American complement of canids—the coyote, the timber wolf, the red wolf, and the now extinct dire wolf of the late Pleistocene. From these analyses models of hominid diversity have been constructed.

REFERENCES

Eldredge, N., & Gould, S. J. Punctuated equilibria: An alternative to phyletic gradualism. In T. J. Schopf (Ed.), *Models in paleobiology*. San Francisco: Freeman, 1972. Pp. 82–115.

Stanley, S. M. A theory of evolution above the species level. *Science,* 1975, *72,* 646–650.

8

Variability and Speciation
in Canids and Hominids[1]

Roberta L. Hall

In North America the genus *Canis* is represented by three species—the wolf *(Canis lupus)*, the coyote *(Canis latrans)*, and the dog *(Canis familiaris)*. (A fourth group, the red wolf *(Canis rufus)*, probably should not be included as a major wolf species; its classification will be discussed in detail later in this chapter.) As field naturalists judge them, these three species are "good" species—that is, their morphology, behavior, and habitats are sufficiently different that the three populations can be considered distinct. However, anatomical, geographical, and behavioral overlap does occur among the three groups and individual members cannot always be classified correctly. Furthermore, genetic differences among the three groups are relatively slight and certainly no obvious differences in chromosomes exist among the three groups. In addition, interbreeding between wolves and coyotes, between coyotes and dogs, and between dogs and wolves produces viable offspring.

Do coyotes, wolves, and dogs belong to three species or to one? Several points of view may be offered. My coeditor, Henry S. Sharp, believes that the North American canids comprise a single population with three different ways of relating to the environment, including domestication by man. In his view, behavioral specializations, rather than genetic or anatomical specializations, differentiate the three groups; hence they should not be considered separate species.

[1] The financial support of the Oregon State University Research Council, through NSF Institutional Grant for Science GU3662, is noted with appreciation.

However, I believe that the three groups clearly correspond to three "good" species as field naturalists judge them and that the niches that the three groups fill may be considered three related modes of adaptation. In most animal lineages, reproductive barriers are largely behavioral but nevertheless real. In the case of a highly intelligent group of animals like the canids, behavioral segregation is doubly important. For the most part, behavioral mechanisms—together with ecological preferences—do keep the three groups apart physically and do keep each separate group functioning as a unit of evolution (i.e., as a species deserving separate status in the taxonomic system).

Regardless of which framework is preferred, it is clear that the North American canids—particularly coyotes and wolves—present an interesting test population in which intergroup variation can be studied profitably. In the canid case we can examine the extent of differences that exist between the living groups and we can trace anatomical separation by studying the remains of ancient animals. The total picture of differences, morphological and behavioral, can be related to an evolutionary framework.

Since we are interested in using canids as analogs to early hominids, our next step is to compare the patterns of variation between the two groups of canids with patterns of variation seen in two groups of fossil hominids. The behavior and physiology of the early members of our own lineage, usually classified in the genus *Australopithecus* or called australopithecines, can be inferred. We cannot observe directly the extent of differences between the groups, which have been extinct for at least a million years. Still, we wonder: How different were they one from the other?

Did they consist of one group of hunters, from whom modern *Homo sapiens* is descended, and one group of vegetarians who left no descendants, as some authors—J. T. Robinson, for instance—suggest? Or were the two groups essentially similar in behavior, differing only in size and to some extent in proportion?

Rather than attempting to settle the issue on the basis of *a priori* reasoning, we are developing a model based on a group of animals whose anatomy, behavior, and range of variation are known.

VARIABILITY AND SPECIATION IN CANIDS AND HOMINIDS

The origin of the mammalian order of the carnivores goes back nearly 65 million years to the early days of the Cenozoic era, the Age of Mammals. But the seven modern families that comprise the carnivore order—

dogs, bears, raccoons, weasels, civets, hyenas, and cats—developed their own distinctive characteristics slowly throughout the era. None of the early Cenozoic carnivores resembles a modern dog, bear, or cat.

Above all else, evolution is a process of experimentation. Animals radiate from a major stock and diversify as they develop new niches, some of which prove viable. But whether individual experiments succeed or fail, the process of experimentation goes on. New types of animals are spawned and new niches are devised. The fate of most experimental groups is extinction, yet some develop relationships with their environment that are both innovative and stable.

The Age of Mammals is divided into five epochs of the Tertiary period: the Paleocene, Eocene, Oligocene, Miocene, and Pliocene (see Table 1). Though the earliest group of true carnivores, known as Miacidae, appeared in the Paleocene, the earliest epoch of the Cenozoic, diversification into separate families was a slow process. Still, by the Oligocene, the third epoch, early members of the canid stock of the *Cynodictis* genus are identifiable. At that time the bears—the living family of carnivores most closely related to the canids—probably still were members of the canid lineage. In the next epoch, the Miocene, the genus of dogs known as *Tomarctus* developed traits that differentiate modern canids from other carnivore families today. These traits include a smaller tail, relatively long feet and legs, a reduced fifth (inner) digit, and body proportions similar to those of modern foxes and wolves. *Tomarctus* had a braincase larger than that of its progenitor *Cynodesmus*, whose brain is larger than its ancestor, *Cynodictis*. The convolutions of the brain increased in complexity, also. The genus *Tomarctus* continued into the Pliocene, and it gave rise to the modern genus, *Canis*, which includes wolves, coyotes, jackals, and dogs (Kurten, 1968; Matthew, 1930).

The fossil record will not allow us to say whether the genus *Canis* evolved in Asia or in North America, for in the later Pliocene it probably was common for animals to pass back and forth across Siberia, the Bering Strait, and Alaska. Biogeographer George Gaylord Simpson (1947) uses the term *Holarctic fauna* to refer to animals that inhabited a vast arctic area of America and Asia, and we should number the early ancestors of the modern wolf among these.

Three species of *Canis* of the early Pleistocene have been described: Falconer's dire wolf, *Canis falconeri;* the Arno dog, *Canis arnensis;* and the Etruscan wolf, *Canis etruscus*. Probably the last-named species served as progenitor for the modern wolf, *Canis lupus*, which is known to have been in Europe during the Günz glacial period, Europe's first major continental glaciation, nearly 1 million years ago. The early wolf was smaller than the modern Near Eastern subspecies of *Canis lupus*. Fossil specimens of the

TABLE 1

Wolf–Coyote Lineage and Geological Sequences[a]

Period	Epoch	Began millions of years ago	Duration in millions of years	Wolf–coyote lineage representative
	Recent	.01	.01	*Canis*
Quaternary	Pleistocene	1.8	1.8	*Canis*
Tertiary	Pliocene	5	3.2	*Canis*
	Miocene	22	17	*Cynodesmus; Tomarctus*
	Oligocene	38	16	*Cynodictis*
	Eocene	54	16	Miacidae
	Paleocene	65	11	Miacidae

[a] Estimates of the antiquity of the geological epochs are from Berggren and Van Couvering, 1974.

genus *Canis* in North America in the early Pleistocene normally are not referred to a species, but are described as wolves of the genus *Canis* with some primitive features. We can assume that during those periods of the Pleistocene when a wide land bridge—a region now referred to as Beringia—connected Asia and North America, the wolves migrated from the Old World to the New, setting a pattern to be repeated in the last period of the Pleistocene by man. Probably the mid-Pleistocene migrants who were members of the species *Canis lupus* superceded the earlier wolf population in North America either by outcompeting the resident wolves or by breeding with them. The fossil evidence of the genus *Canis* in the very early period of the Pleistocene is so scanty that neither argument can be made very forcefully.

The modern golden jackal, *Canis aureus*, is known only in the Old World today, and it appeared in the late Pleistocene. However, other species of jackal have been reported in Africa as early as the Villafranchian. (*Villafranchian fauna* refers to a cluster of animals that characterized Old World faunas in the late Pliocene and early Pleistocene.) Though the jackal is considered the ecological analog to the New World coyote, the two are not genetically related. Rather, each probably evolved from basic wolf stock, one in the Old World and one in the New.

In addition to the coyote and the wolf, a third species, the dire wolf, *Canis dirus*, lived in North America in the late Pleistocene. This wolf became extinct approximately 8500 years ago, about the same time that many other animals, including the saber-toothed cat and many species of large herbivores, became extinct. The significance of the dire wolf will be discussed in the following chapter.

PLEISTOCENE GLACIATIONS AND COYOTE EVOLUTION

By alternately creating and eradicating geographic barriers, the continental glaciations of the Pleistocene profoundly affected the evolution of North American animals. The locking up of ice in glacial sheets lowered the level of the seas and created Beringia, the land that connected Asia and North America. Sometimes ice sheets divided North America into a southern zone and a northern Holarctic zone, including Beringia. Between these icefree areas migration was difficult and sometimes impossible for extensive periods (see Figure 1).

In the southern zone, during one of the glacial periods, it is postulated that the basic wolf population evolved rapidly into ancestral coyote stock. When the ice barriers receded and corridors between the southern and northern zones opened, the two wolf groups did not fuse but continued to evolve separately in slightly different directions. There remains a question of dating, for we would like to have the answers to two questions: Did the coyote differentiate from the wolf before or after *Canis lupus,* as we know it, developed? For approximately how long have the two groups, coyote and wolf, been separated?

In attempting to identify the period during which wolves and coyotes diverged we must turn to the fossil record, which at worst is ambiguous and at best is difficult to interpret. The evolutionary history of no American Pleistocene mammal is known in detail, but the scientific literature does contain many references to members of the genus *Canis.* These data can be used to develop a model for the evolution of the coyote. A study of the fossil evidence indicates that the coyote has existed at least since the most recent glacial period, the Wisconsin, but probably did not exist before the penultimate major glaciation, the Illinoian. Thus the wolf population evidently separated during the Illinoian period, a separation that fostered evolution of the coyote, *Canis latrans.* Under this hypothesis, we suggest that the coyote developed out of the modern wolf species approximately 500,000 years ago.

Until recently, no global model of climatic change or of faunal evolutionary sequences existed. Yet climatic changes—and in particular climatic deterioration or general cooling trends—have characterized earth history of the past 15 million years, and the trend has accelerated during the past 3 million years. Indirect and direct effects of climatic change on the evolution of specific groups of plants and animals have been far-reaching. Consider, for example, the major continental glaciations. These glacial periods altered environments and changed the selective forces working on all the animal species. Sea levels lowered and land bridges appeared, linking islands with continents and permitting migration be-

Figure 1. Land bridges and glacial barriers in the Ice Age. The polar-projection map shows approximate continental relationships 18,000 years ago at the height of the Wisconsin glaciation; a similar pattern probably existed during the Illinoian glaciation of half a million years ago. A portion of modern-day Alaska and much of the present Bering Sea were an integral part of the largely Asian tundra; it was there that the wolf, *Canis lupus,* was at home. The coyote, *Canis latrans,* evolved from wolf stock south of the glacial barrier. (Tundra is indicated by horizontal lines; sea ice, by cross-hatching; and continental ice, by diagonal lines.)

tween continents. Across Beringia many animals made their way—among them members of the genera *Canis* and *Homo*.

Because of the obvious interrelationships that exist between radically changing climates and the evolution of animal forms in various parts of the world, the publication of a system that unifies information on ancient climates and life forms has aroused the interest of specialists in many fields. The system referred to and followed here was developed by W. A. Berggren and J. A. Van Couvering and was published in 1974. Table 1 uses the chronology established by Berggren and Van Couvering to chart the phylogeny of canids throughout the Cenozoic, the Age of Mammals; Figure 2 provides the basic chronology of faunal and glacial sequences in North America.

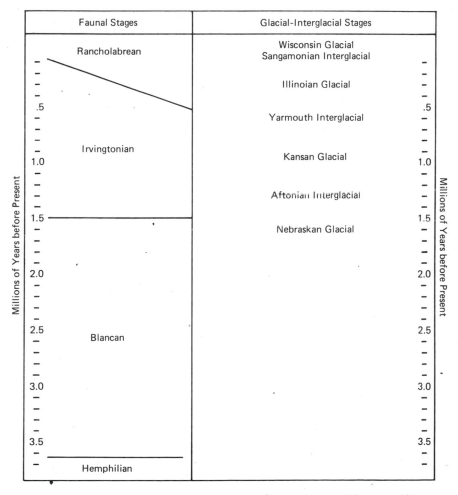

Figure 2. Correlation between North American faunal and glacial stages. (Adapted from W. A. Berggren & J. A. Van Couvering, 1974, p. 144.)

One of the many problems that plague geologists and paleontologists working in areas that have been glaciated is that later glaciers remove the evidence of earlier periods and often confound attempts to determine the geological age of specimens of fossils that have been found. Therefore, some of our best fossil data during the glacial phases come from Alaska and from the southern half of the United States—areas that were not covered by glaciers. Most of the fossil canids are dated geologically; that is, in relation to particular strata or to a major climatic event such as a glaciation, or as members of a cluster of fauna with a defined antiquity. Reference to Figure 2 will help in interpreting the data on canid fossils.

Remains of canids positively identified as coyotes have been found in deposits of Wisconsin age in California, Texas, Idaho, and Florida with a pre-Wisconsin but post-Illinoian date noted in Maryland. The majority of coyote fossils have been found in faunal clusters of the Rancholabrean group but some have been noted in the long Irvingtonian faunal period, which may have begun as early as 1.4 million years ago and hence may include both the Kansan and Illinoian glaciations as well as the periods following them (see Figure 2). The extent of the Irvingtonian makes it difficult to provide an estimate of the antiquity of these specimens. Conservatively, the paleontological record gives unambiguous evidence of the existence of the coyote only in post-Illinoian deposits (Dalquest, 1965; Gidley & Gazin, 1938; Johnston & Savage, 1955; Kurten & Anderson, 1970; Ray, 1958; Savage, 1951; Schultz, 1938; Slaughter, 1966; Stock, 1953). Applying the model of geographic speciation, we can hypothesize that the coyote evolved during a phase of the Illinoian glaciation and that the wolf and coyote have represented separate species for at least 500,000 years.

How long does it take for two populations to become separate species? No consensus exists among theorists on this question, but it is generally agreed that the rate of speciation depends on the rapidity of the environmental change to which a breakaway population is adapting. In the case of the coyote, I suggest that a population of wolves living at the southern periphery of the Holarctic *Canis lupus* range was cut off by an ice sheet and was confined to an environment different from that to which the parent group had adapted successfully. Glacial conditions altered the habitat of the ancestral coyotes and pushed them farther south, opening up new challenges and new possibilities for them.

Under these circumstances evolution could proceed quite rapidly and perhaps produce a population that within 10,000 years field naturalists would judge to be a "good" species. After the ice sheet retreated, ecological and behavioral features—differences in habitat preferences and mutual antipathy—probably kept the wolf and the coyote populations evolving separately. Occasional hybridizing between the two groups would not have been sufficient to shatter either of the two carnivores' holds on its own ecological niche.

VARIATION IN MODERN CANIDS

I have developed a model for the evolution of the coyote *Canis latrans* out of the basic wolf population in North America. The advantage of studying the canids as a model for early man is that we can analyze the behavior and anatomy of the contemporary, descendant populations, and hence can relate behavior and appearance of living animals to fossils. This discussion of the variability of modern coyote and wolf will include the red wolf, *Canis rufus*,[2] a possible third species of wild canid in North America today, and will also consider the New England "coy-wolf," a new population of wild canids that became established in New England in the late 1950s.

The coyote and wolf are considered "good" species even though they are interfertile and both have been known to mate with domestic dogs and produce viable offspring. Generally, evolutionary biologists do not consider that interfertility affects the definition of a species if the mating occurs due to human intervention, whether in confinement or due to ecological disturbance. The critical question in determining whether a population represents a species is this: Does the population represent a unit of evolution within which evolutionary pressures, such as selection and mutation, are felt and expressed? If hybridization is rare, it will not affect the genetic integrity of the group or its evolutionary direction and hence should not affect species definition.

The most obvious difference between wolf and coyote is size. Adult coyotes normally range from 23 to 45 pounds, whereas adult wolves average about two and one-half times as large; in most regions wolves range from 60 to 100 pounds. But there are differences in shape as well. The coyote has a comparatively longer, leaner muzzle and lighter jaw structure (Figure 3). Differences of proportion as well as of size can be related to, and perhaps explained by, functional requirements for killing and eating prey of quite different sizes. Moose, bison, and musk-ox are the largest prey taken by wolves, with deer or caribou being more common prey in some areas. Coyotes feed mainly on rodents (mice) and lagomorphs (rabbits) and only occasionally tackle prey as large as deer or

---

[2] *Canis rufus*, the red wolf, used to be known formally as *Canis niger*; Gipson, Sealander, and Dunn (1974) presented the name change.

Figure 3. (a) Wolf muzzle. Wolves tend to have relatively short and broad muzzles compared to those of coyotes. (Photograph by Don Alan Hall.) (b) Coyote face. (Photograph by Roberta L. Hall.)

domestic sheep. Where small mammals are not abundant, coyotes consume insects and plants and are adaptable in meeting their dietary needs.

The same kinds of measurements that differentiate the wolf from the coyote differentiate the large hominid *Australopithecus robustus* from the smaller, more delicate *Australopithecus africanus*. According to Barbara Lawrence and William Bossert (1967), "Wolves have a relatively small brain case and massive rostrum (muzzle). The latter presumably is a reflection of the large size of the animals on which they prey [p. 224]." The measurements Lawrence and Bossert found most critical in distinguishing the two canid species reflect the much greater width of the wolf's face and the greater size of the wolf's teeth. The wolf also has a well-developed sagittal crest, indicating a large temporal muscle for operating the jaw.

All these traits—relatively smaller braincase, relatively well-developed sagittal crest, relatively wide face, and massive teeth—have been used to differentiate the large form of australopithecine from the small form (Figure 4): "Much of the difference in general skull form between [robust form] and [gracile form] is, therefore, due to the differences in nature and amount of use to which the dental battery is put [Robinson, 1954, p. 331]."

The strikingly similar kinds of differentiation between the two forms of early man and the two forms of *Canis* cannot be used to suggest that the dietary habits of the early hominids were identical to those of the modern canids. Both groups of animals are omnivorous, but coyotes and wolves rely on meat for approximately 90% of their diet and the earliest hominids may have relied on vegetable matter for 50% of their diets. The important

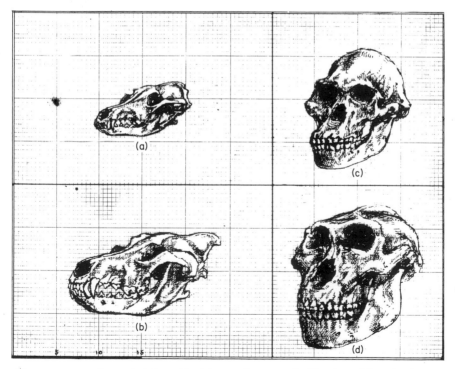

Figure 4. (a) Coyote, (b) wolf, (c) gracile australopithecine, (d) robust australopithecine. (Drawing by John Slater.)

point is that wolves and coyotes eat the same kinds of food, but the large masticatory apparatus of the wolf is adapted for eating more food in larger packages. It is possible that the same kind of model works for the early hominids (Cachel, 1975). The larger form may have eaten the same kind of food that the smaller form ate, but in larger packages and in larger quantities.

Wolves tend to hunt in groups, probably because of the size of their prey, whereas coyotes often hunt alone. It has been assumed by many authors that, because wolves favor group hunting and coyotes are often seen hunting alone, the wolf has a well-developed social life and the coyote lacks one. Studies of coyote social life reveal that in some situations coyotes maintain social bonds, probably of an extended-family nature, that focus on the rearing of young (Ryden, 1975). Social differences are of degree, not kind.

Differences in prey size may not affect social requisites, but they do appear to affect at least two important aspects of ecology and behavior. First, because wolves prey on large animals they tend to have a large geographic range. Second, the wolf's size, greater visibility, and tendency to hunt large animals in packs—together with its traditional role as evil genius in Western folklore—make the wolf much more incompatible with farming and with North American farmers. With the spread of farming across North America, wolves have been exterminated and coyotes have moved into large areas where they were not found aboriginally.

Like most other writers who have compared wolves and coyotes, I have discussed effects of size differences on various aspects of the two species' ecological niches. If I wanted to explain the size differences, though, I could have argued that the existence of the potential for two different niches brought about the differences in size and behavior. Surely there existed in aboriginal North America ecological space for two related but different wild canids—coyotes in semiopen brush country and wolves in the timbered and open northern areas. Anatomical differences were selected for, and the two groups did not interbreed to any extent in border areas because of behavioral differences. The postulated controlling factor is ecology—behavior enforced the ecological imperatives but did not create them. Localized hybridization that might occur with the breakdown of the ecological and/or behavioral barriers would not affect the overall distinctiveness of the two populations.

The most important evidence challenging the concept that wolves and coyotes are distinct species is the existence of the red wolf of the southern states and a wild canid in New England that first appeared in the late 1950s. Coyote–wolf interaction is known generally to be hostile; hence, the possibility of there existing hybrids between the two species was at first considered quite unlikely. But recent developments concerning these two intermediate-size wild canids have changed zoological opinion.

The aboriginal range of the red wolf was the southern portion of what is now the United States from Texas east to Florida. The red wolf was exterminated from much of its eastern territory and in the mid-twentieth century a number of studies were undertaken in its western range to determine whether the red wolf was a separate species. One of the factors promoting the studies was the observation that in Texas, where the western extremity of the red wolf's range met the eastern extremity of the coyote's range, the two groups appeared little different in size. This was not expected, since closely related species normally differ most in overlapping zones—theoretically in order to reduce competition between them and permit survival of both. Hence, hybridization along the border area was suspected. Multivariate tests made using size and proportion measurements of wolf, red wolf, coyote, and dog populations led to the conclusion that the red wolf, presently a rare animal, should be considered a geographic subspecies of the wolf *Canis lupus,* and one that, under severe attack by man, hybridized with the coyote. [But see Paradiso (1968) and Paradiso and Nowak (1973) for an alternative opinion.]

If the modern red wolf should be considered chiefly a wolf but with some coyote admixture, a second canid that also appears intermediate in size between the two species may be considered largely coyote, with some genetic input from the wolf. In the late 1950s, this large yet coyote-like wild canid entered New England, filling a void left when the wolf was exterminated there at the beginning of the twentieth century. At first considered to be "coy–dogs"—animals resulting from coyote–dog hybridization—these wild New England canids are now judged to have coyote and wolf ancestry. This conclusion was reached on the basis of studies of laboratory-bred hybrids and behavioral and metrical studies of hybrids of the wild canids themselves (Lawrence & Bossert, 1969, 1975). One of the factors working against the coy–dog hypothesis is that the breeding season of coyote–dog hybrids is too early in the year to ensure survival of young in the wild. Another factor is that coy–dog males, reared in the laboratory by Helenette and Walter Silver (1969), behaved as domestic dogs and did not help care for their young. In contrast, the New England canids had the well-organized family rearing pattern common to wild canids. Still, there is some question as to whether the Silvers' study of hybrid coy–dog males is conclusive, since the coy–dogs were reared in a human environment. If coy–dogs were reared in the wild, they might behave more like wild canids.

It has been postulated that the normal antipathy of the wolf to the coyote broke down in New England when isolated wolves found only coyotes with whom to mate. The origin of the particular ancestors of the New England canids is not known. Stories explaining their introduction range from tales that pet wolves kept by local persons became free, to suggestions that a few immigrant wolves came in from Canada and met a few coyotes coming east. In either case it is assumed that because both groups were small in numbers their normal antipathy broke down.

The new canid population possesses some features of both wolf and coyote, though the size range is intermediate. Some individuals look like pure wolves or coyotes. Considering genetic models, it seems reasonable to hypothesize the control of metric traits by many genes, each having a small effect, and control of morphological traits by single genes. Observed features of the New England canids fit such a hypothesis if hybridization was recent.

It seems clear that with the New England hybrid we have evidence— even better evidence than in the case of the Texas red wolf—that hybridization is still possible in the wild between coyotes and wolves. It also seems clear that ecological and behavioral factors rather than genetic ones are responsible for the usual maintenance of separation between the two species. If the New England canids are allowed to survive and become established as a species, it will be interesting to observe whether the morphological–behavioral features remain intermediate, or whether the group will gravitate to the form and niche of one of the parent populations. Is there ecological space for an intermediate-size wild canid?

It is possible and perhaps likely that intermediate forms have arisen in the past only to be overcome and absorbed by one of the parent groups. Two issues are involved: the genetic issue of interbreeding and the ecological theory of the niche. Since interfertility of the species exists in areas of overlap, only behavioral mechanisms keep wolves and coyotes apart. But anatomically, as well as ecologically and behaviorally, the coyote and the wolf are adapted to prey on animals of somewhat different sizes in somewhat different environments, and it is the ecological factor that must be considered determinant in the maintenance of the two populations.

Where does an intermediate canid fit in? Clearly man is implicated in the generation of the New England hybrid—by exterminating the wolf and probably by letting captive coyotes loose to become parents of the new populations. Did human cultural activity—farming—also promote hybridization of the two species by offering an opportunity for an intermediate-sized canid predator?

My interpretation of the meaning of species in reference to wolf, *Canis lupus,* and coyote, *Canis latrans,* has an ecological–behavioral base and, I believe, represents a healthy shift from the rigid definition of a species as a pure genetic isolate. This approach to the definition of species also would benefit the study of early hominid populations. As noted in the Introduction, many ecological and behavioral factors suggest the study of North American canids as analogs to hominids; study of the dynamics of wolf–coyote interaction should provide a useful model for the investigation of speciation in early hominids. A review of the problem of species definition in early hominids illustrates this point.

AUSTRALOPITHECINE VARIATION

Basically, the problem in defining the australopithecine populations is that many incomplete fossils of early hominids have been found, yet it is unclear how many species are represented by these finds. As members of a species studying itself, we are curious to determine which of the hominid groups (if any) we should consider as our own ancestors. We are also curious to discover what pressures led to the emergence of our species.

One feature of *Homo sapiens* that differentiates it from other species is that despite a wide geographic distribution and adaptation to many environments, no important differences exist between separated populations, and no negative results follow interbreeding between populations. Some authorities argue that ever since the origin of culture humanity has been a single species. Others perceive several species as late as the early part of the Wisconsin glaciation. Anthropologists have generally attempted to answer the question through the examination of fossils. They have attempted to test whether anatomical traits cluster into two or more groups, indicating multiple species, or whether variability is random, indicating one variable species.

As Sherwood Washburn noted in addressing a special session on early man at the April, 1972 conference of the American Association of Physical Anthropologists at the University of Kansas, the greatest consensus regarding the taxonomy of early man exists where the least evidence is found. He was referring specifically to the general agreement that *Ramapithecus*, an apelike creature that lived 8 to 14 million years ago, was the forebear of *Australopithecus*. But the general concept holds for all aspects of the study of early man, and no doubt for fossil studies of other lineages as well. When the discovery of the gracile *A. africanus* was followed by the finding of a more robust hominid, now known as *A. robustus*, it seemed clear that the gracile species was carnivorous, or at least omnivorous, and led, slowly, to *Homo*, whereas the robust form was a vegetarian that became extinct. In recent years, fossil hominids have been found in greater abundance and the new data have added complexity to the problem. No longer does the simple progression of *A. africanus* to *Homo* seem so obvious.

Unfortunately, the issue often becomes merely a philosophical one, with the argument distilling to one between "splitters" and "lumpers." In studying fossil populations, we cannot make direct observations to determine whether a species is "good" or not. Therefore investigators often believe they are forced to take a philosophical position on whether to err on the side of naming too few species or on the side of naming too many. Splitters tend to name a large number of species, considering new finds separate lineages until they are demonstrated to be similar to previously named species. Lumpers place new finds in an established category until they are proven not to fit neatly. Because the data are almost always fragmentary, once a philosophical position is taken the evidence can be manipulated to support it. Despite the uncovering of much new data on early hominids, the positions regarding splitting and lumping changed little in the 1960s and 1970s.

We can use three investigators to illustrate the splitting and lumping traditions. C. Loring Brace (1967) can be considered a lumper and Louis and Richard Leakey extreme splitters. Brace has suggested, as one possibility for further study, that the two australopithecine groups are sexually differentiated, with the robust form being male and the gracile form being female. In any case, he does not think that the evidence demonstrates the existence of more than one species of early hominid. Louis Leakey and his son Richard have argued that the genus *Homo* existed coeval with *Australopithecus*, and that australopithecines represent an evolutionary backwater unrelated to modern man. Their conclusions about the definition of the genus *Homo* raise unresolved problems.

One of the most useful concepts in the study of the evolutionary history of organic forms is the concept of adaptive radiation. Following a population's breakthrough into a new adaptive zone, whether by a change in behavior or a change in environment—or both—a rapid diversification or adaptive radiation occurs at all levels of the taxonomic hierarchy. Examples include the diversification of early reptiles on land, lemurs on Madagascar, apes in the Old World during the Miocene epoch, and, possibly, hominids in Africa in the early Pliocene and canids in Holarctic and temperate zones in the Pleistocene. Following the radiation of new types to fill available niches in a new zone, the rate of evolutionary change again relaxes and a stability of forms prevails. Niles Eldredge and Stephen Gould (1972) have proposed a model of punctuated equilibrium to explain diversification. They suggest that diversification occurs on the periphery of a group's range; when new forms first arise they differ anatomically at least as much as they will differ later.

The basis of their expectation is clear. If separation of two groups is to succeed, the groups must occupy different niches. During the period in which genetic incompatibility is evolving, separation is most likely to be achieved if anatomical differences are marked. It is assumed that the anatomical differences reflect ecological and ethological differences that, unfortunately, are not preserved in the fossil record.

In the 1970s, estimates of the antiquity of early hominids were extended back from the approximately 1 million years accorded Raymond Dart's original find, *A. africanus,* to possibly 5.5 million years suggested by finds at Lothagam in East Africa. Though at this writing analysis of the material is not complete, at least two hominid forms appear to go back at least to 4 million years ago. As in the wolf–coyote model, the endurance of two (or more) forms does not indicate that no interbreeding occurred. Rather, it indicates that any interbreeding that did occur was not sufficient to shake the ecological stability of the two groups.

It is not necessary to accept the hypothesis, offered by J. T. Robinson (1954), that the ecological separation of the two hominids involved fundamental differences in choice of food, with the robust form being vegetarian and the gracile form eating meat. It is more probable that both hominids were omnivorous, but that the size and kind of prey and the type of hunting behavior differed between them. We have seen that on the average wolves are about $2\frac{1}{2}$ times heavier than coyotes. Estimates of weight differences between *A. africanus* and *A. robustus* vary; most investigators suggest that the robust form was at least $1\frac{1}{2}$ times the weight of the gracile form. It has been suggested that an average member of the gracile form probably weighed about 70 pounds; a member of the robust form probably weighed about 95 pounds (Pilbeam & Gould, 1974). These estimates place both australopithecines below the lower weight range of modern man. C. Owen Lovejoy and Kingsbury G. Heiple (1970), however, suggest a weight of 40 to 50 pounds for some of the gracile australopithecines. Clearly no conclusions exist regarding the extent of size differences, but we may be safe in assuming that the robust form was perhaps $1\frac{1}{2}$ times the weight of the gracile form and had a substantially greater crushing power available in its jaw.

Most carnivores today are opportunistic, taking whatever food is most available, culling the weak and young, and scavenging when necessary or beneficial. Australopithecines probably followed this behavior pattern, but no doubt they had unique, primate-specific techniques of killing game (Figure 5). These early hominids may have used some of the techniques of deception chimpanzees use in preying upon small animals, as described by Geza Teleki (1973).

Perhaps the smaller form of hominid hunted alone, as coyotes usually

Figure 5. Australopithecines hunting. (Drawing by John Slater.)

do, and preyed on small animals. The coyote still practices a full social life involving food sharing and care of young and it is possible that *A. africanus* did too. It may be postulated that, like the wolf, the larger form of hominid obtained a significant portion of food by cooperative hunting. Traditional forms of analysis of fossil hominids have overlooked this hypothesis. Instead they have concentrated on the size of the australopithecine molars, assuming that molar size is correlated to the degree of dependence on plant food. Traditional analysis has also noted that typical specimens of *A. robustus* have proportionately small front teeth, and it has been inferred that the relatively large front teeth of *A. africanus* indicate a tendency to eat meat. But it has never been established that australopithecines used front teeth to kill game or to tear meat. Indeed, it seems unlikely that large front teeth would have been selected for in an omnivorous primate, because hands and tools probably were used to kill game and to divide it. It seems certain that relatively small front teeth do not conclusively indicate vegetarian subsistence. In fact, small front teeth would prevent the australopithecines from harvesting some kinds of food used by modern gorillas, who husk bark and bamboo with their anterior teeth.

   As with the wolf and coyote, it is probable that the ranges of the different forms of hominids overlapped little. Species borders may have shifted back and forth over the years, however, perhaps in line with microevolutionary environmental changes. As appears to be the case with wild canids, behavioral mechanisms probably enforced separation of the two species, with hostilities occurring when avoidance mechanisms failed. George Schaller (1972) noted that carnivores in Africa occasionally regard one another as prey. Usually, however, they see other carnivores as competitors to be avoided or chased, not to be eaten. When one considers early hominids as competitors to other scavengers and predators, rather than as usual prey, one removes some of the difficulties anthropologists have had in postulating elaborate and undemonstrated defense mechanisms for them.

THE GENUS *HOMO* AND TWO TAXONOMIC MODELS

Related to the taxonomic problem of interpreting anatomical differences in australopithecines is the very serious problem of defining the genus *Homo*. Following Wilfred Le Gros Clark's (1964) definition, anthropologists have identified *Homo* on the basis of a number of anatomical traits derived from a listing of the features found in specimens classified as members of that genus. Clark's definition does not include specification of ecological or cultural properties, nor does it tackle the problem of whether individual fragments are to be considered in the genus *Homo* if they possess one or more of the requisite morphological traits. Yet most anthropologists agree that *Homo* is culturally and ecologically distinct from *Australopithecus*, and assume, specifically, that *Homo* is characterized by highly developed big-game hunting techniques, by the tendency to occupy cold as well as warm climatic zones, and by the use of fire in cold latitudes. Glynn Isaac discussed this problem in a paper given at the 1975 session of the American Association of Physical Anthropologists. He said, "I offer as a working hypothesis the suggestion that in the archaeological record of $2\frac{1}{2}$ to 1 million years ago we are seeing traces of a nonhuman adaptive system that nonetheless incorporated the evolutionary foundations of humanness [Isaac, 1976, p. 33]." Isaac's working hypothesis underlies the hypothesis of this book, that the behavior system of the early hominids has much in common with certain extant nonhuman and nonprimate social carnivores. The peculiarly human use of resources and the peculiarly human way of conceptualizing probably did not appear until, at most, a million years ago—markers for this change include the controlled use of fire, as well as a more sophisticated tool-kit. One other marker may be offered, the extinction of all but one hominid form—this indicates a widening of the hominid niche.

George Gaylord Simpson (1963) has reminded anthropologists that a genus is a different kind of category from a species. The extent of differences between animal populations that warrant generic distinction is imprecise and subject to the investigator's good judgment. However, it is generally held that generic distinctions must rest not merely on morphological distinctiveness of populations but also on the occupation of different ecological plateaus. In this way, Robinson (1954) argued, the two australopithecine groups should be labeled distinct genera because he believed they represented different adaptations—*A. robustus* subsisting on vegetable matter and *A. africanus* on a diet of meat. The evidence, however, does not support this inference.

First Model. Let us follow the wolf–coyote model and postulate a difference only in size of prey and in pattern of hunting. In the absence of evidence that any early hominid population inhabited an ecological plateau different from that inferred from australopithecine archeological deposits, we should question whether any fragmentary remains legitimately can be considered in the genus *Homo*. It is too soon to argue that any particular group of australopithecine did not give rise to the genus *Homo* and at the same time too late for any investigator to claim to have discovered the exclusive ancestor of modern man. We know too much of the complexity of the problem for either of these claims to be taken seriously.

Let us assume that two species of hominids coexisted in Africa for several million years, each avoiding competition with the other by seeking different kinds of prey and avoiding the other's territories. Let us also assume that hybridizing occurred too infrequently to affect their stability. Maintenance of two species of hominids could continue only as long as the niches of each remained relatively narrow. However, with the evolution of *Homo*, probably in some peripheral population of australopithecines, the niche broadened and absorbed the ecological styles of both hominids.

Specifically, the rise of the genus *Homo* must have spelled extinction for other hominids. Following Eldredge and Gould's model of punctuated equilibrium, it is reasonable to postulate that the distinctive features of the genus evolved relatively rapidly. Though we may find that some individual members of the early hominid populations foreshadowed anatomical features to be developed more fully in *Homo*, we should not, without more evidence than has come forward so far, take seriously the concept that one or more species of *Homo* coexisted with several species of *Australopithecus* for a long period. From an ecological–ethological point of view, such coexistence is not likely. In contrast, the coexistence of two or more australopithecine species is comparable to the pattern seen in coyote–wolf relationships.

Second Model.   We may refer to the foregoing as the *strict* or *conserva-tive model of hominid taxonomy*—it deserves this designation because of its insistence on ecobehavioral criteria for discriminating between genera. However, since an increasing number of fossil finds from the Plio-Pleistocene period, 3 million years ago, are being classified as morphological in the *Homo* genus, it may be that our nomenclature should reflect these facts. Though Louis Leakey and J. T. Robinson have sup-ported this proposition for years, the consensus did not change in their direction until multinational research teams working in East Africa in the late 1960s and early 1970s found numerous, well-dated fossil fragments that appeared to resemble the genus *Homo* more than the genus *Aus-tralopithecus*. The two traits involved are cranial morphology (size and shape), and jaw proportions. The skeletal remains are fragmentary and reconstructions tentative; hence it has required numerous specimens to convince scholars that there might have been an early population—not just a few aberrant specimens—that should be classified as the genus *Homo* (Jolly & Plog, 1976). Since biological taxonomy is a tool of science rather than a law of nature, there exists an alternative way to handle the designation *Homo*.

Clearly the broad human econiche including fire using, big-game hunt-ing, and intense competition with other predators did not exist until the mid-Pleistocene—about a million years ago. It appears, however, that this *cultural* pattern evolved within a biological species that has a great *morphological* antiquity. We can recognize the antiquity of the morphological *Homo* at the same time that we acknowledge that cultural development proceeded without immediate biological effects on the culture-bearing species.

One advantage offered by this approach is the insight that culture-dependence is a relative phenomenon that varies along a continuum. This perspective helps us to be more keenly appreciative of the cultural behav-ior of other living species; among other advantages, it helps to bridge the gap from wolf culture to hominid culture.

To recognize the morphological *Homo* in late Pliocene or early Pleis-tocene deposits implies that both *Australopithecus* and *Homo* were prod-ucts of a hominid radiation prompted by the opening up of new savannah environments. These forms probably coexisted in niches no broader—and possibly somewhat narrower—than those occupied by modern canid species. Coexistence of the several hominid species continued until one of the forms, *Homo*, developed a cultural proficiency that brought about an increase in its population, geographic dispersal, and ecological breadth. These factors led indirectly but inexorably to the extinction of the other hominid forms (Figure 6). In other mammalian lineages, adaptive radia-

Figure 6. The magic of fire. In this fanciful drawing, one group of early hominids *(Homo)* gathers around a fire while another band *(Australopithecus)* observes in wonder. Coexistence of two or more hominid species would not continue long after one developed unique cultural proficiencies—such as the possession of fire. (Drawing by John Slater.)

tion has worked as a kind of grand evolutionary experiment with forms being "thrown off" on a trial basis and species selection removing those that prove least fit. We should not be surprised to find evidence of the same processes occurring in early hominids. Unfortunately, at the present time we cannot define the ecological barriers that separated the various branches of the hominid radiation, or that permitted them to coexist for several million years. North American paleontologists have found skeletal remains of *Canis latrans* and *Canis lupus* in the same site, yet we know that coyote and wolf do not normally share the same territory at exactly the same time. Probably the same kind of effect occurs in the hominid fossil record: The several hominid forms did not interact as individuals or share territories. Yet their ranges changed from year to year, so individuals of different species were deposited in the same site within the same geological period.

The only break with biological nomenclature in the second model is that a group is being labeled *Homo* even though it has more in common ecobehaviorally with another genus *(Australopithecus)* of its own time period than it has with later *Homo*. The justification for doing this, however, is that the ecobehavioral adaptation of these hominids was in large part nonbiological. Furthermore, we have modern examples to indicate the necessity for this deviation from nomenclatural rules. For instance, one geographic area, the Philippines, includes human populations that have diverse cultural–ecological adaptations. These populations range from the Tasaday, a hunting–gathering people who exert minimal impact on the environment, to fully industrial people who, like most of the European and Northern American populations, are churning the earth. Clearly, an analysis of the skeletal remains of any Philippine peoples tells us nothing about their cultural adaptations. So it may be for the earliest *Homo* populations: Behavioral inferences must be based on archeological remains, not on skeletal material. Though the anatomical evolution of the hominids occurred in leaps and bounds, according to the model of punctuated equilibrium, cultural evolution proceeded slowly, and at first quite tentatively. The genus *Homo* was slow in recognizing its own power, or the power of the culture it bears.

CONCLUDING REMARKS

In this chapter I have tried to focus on ecological and evolutionary processes operating in canids and in hominids, rather than on nomenclature as such. Labels for canids and hominids change, both with new data and with the intellectual fashion of the times. Whatever the labels used, the types of adaptive niches occupied by the early hominids and by the modern canids have much in common. For this reason we may expect that the pattern of variation and niche differentiation that operates within canids is broadly analogous to the pattern of variation in early hominids.

Classification of fossil hominids has always provoked great controversies simply because, as members of a species studying its own development, we all bring a certain keen, personal interest to our work; we are searching for our *own* ancestors as well as for scientific credibility. For this reason, in the last section of this chapter, I have offered two widely divergent classification schemes, each of which makes equal scientific sense; I must leave it to the reader to choose between them. It is well for all of us to remember that biological classification is a tool of science, not a law of nature. It should be used in that very human process of science—to provoke our intellect and to enrich our perceptions.

Probably the most important asset to be gained from the study of nonhuman evolution is a sense of perspective and proportion. By a close examination of the literature on other animal groups—in this case members of the genus *Canis*—it is possible to develop an *almost* impersonal sense of evolutionary process. The application of this sense of process to the study of the evolution of the genus *Homo* may not provide instant solutions to the problems of human evolution, but it at least suggests new models and stimulates new questions.

ACKNOWLEDGMENTS

I would like to acknowledge the help of my colleagues at Oregon State University: Thomy Smith, for drawing the map of Pleistocene land bridges; Thomas Hogg, for his encouragement; Andrea Campbell and Linda Morgan, for their help in preparing the manuscript; and Richard Darsie, for his help in preparing the subject index.

REFERENCES

Berggren, W. A., & Van Couvering, J. A. The late Neogene: Biostratigraphy, geochronology, and palaeoclimatology of the last 15 million years in marine and continental sequences. *Palaeography, Palaeoclimatology, Palaeoecology,* 1974, *16,* 1–216.

Brace, C. L. *The stages of human evolution.* Englewood Cliffs, New Jersey: Prentice-Hall, 1967.

Cachel, S. A New view of speciation in *Australopithecus* and other early Hominidae. In R. H. Tuttle (Ed.), *Paleoanthropology.* The Hague: Mouton, 1975. Pp. 183–201.

Clark, W. E. Le Gros. *The fossil evidence for human evolution.* Chicago: Univ. of Chicago Press, 1964.

Dalquest, W. W. New Pleistocene formation and local fauna from Hardeman County, Texas. *Journal of Paleontology,* 1965, *39,* 63–79.

Eldredge, N., & Gould, S. J. Punctuated equilibria: An alternative to phyletic gradualism. In T. J. Schopf (Ed.), *Models in paleobiology.* San Francisco: Freeman, 1972, Pp. 82–115.

Gidley, J. W., & Gazin, C. L. The Pleistocene vertebrate fauna from Cumberland Cave, Maryland. *U.S. National Museum Bulletin,* 1938, *171,* 1–99.

Gipson, P. S., Sealander, J. A. & Dunn, J. E. The taxonomic status of wild *Canis* in Arkansas. *Systematic Zoology,* 1974, *23,* 1–11.

Isaac, G. Early hominids in action: A commentary on the contribution of archeology to understanding the fossil record in East Africa. *Yearbook of Physical Anthropology,* 1976, *19,* 19–35.

Johnston, C. S., & Savage, D. E. A survey of various late Cenozoic vertebrate faunas of the panhandle of Texas. Part I: Introduction, description of localities, preliminary faunal lists. *University of California Publications in Geological Sciences,* 1955, *31,* 27–49.

Jolly, C. J., & Plog, F. *Physical anthropology and archeology.* New York: Knopf, 1976.

Kurten, Björn. *Pleistocene mammals of Europe.* Chicago: Aldine, 1968.

Kurten, B., & Anderson, E. The sediments and fauna of Jaguar Cave: II—The fauna. *Tebiwa Journal of Idaho State University Museum,* 1970, *15*(1), 21–45.

Lawrence, B., & Bossert, W. H. Multiple character analysis of *Canis lupus, latrans,* and *familiaris* with a discussion of the relationships of *Canis niger. American Zoologist,* 1967, *7*(2), 223–232.

Lawrence, B., & Bossert, W. H. The cranial evidence for hybridization in New England *Canis. Breviora,* 1969, *330,* 1–13.

Lawrence, B., & Bossert, W. H. Relationships of North American *Canis* shown by a multiple character analysis of selected populations In M. W. Fox (Ed.), *The wild canids.* New York: Van Nostrand-Reinhold, 1975. Pp. 73–86.

Lovejoy, C. O., & Heiple, K. C. A reconstruction of the femur of *Australopithecus africanus. American Journal of Physical Anthropology,* 1970, *32*(1), 33–40.

Matthew, W. D. The phylogeny of dogs. *Journal of Mammalogy,* 1930, *11,* 117–138.

Paradiso, J. L. Canids recently collected in east Texas, with comments on the taxonomy of the red wolf. *American Midland Naturalist,* 1968, *80*(2), 529–534.

Paradiso, J. L., & Nowak, R. M. New data on the red wolf in Alabama. *Journal of Mammalogy,* 1973, *54*(2), 506–509.

Pilbeam, D., & Gould, S. J. Size and scaling in human evolution. *Science,* 1974, *186,* 892–901.

Ray, C. E. Additions to the Pleistocene mammalian fauna from Melbourne, Florida. *Bulletin of the Museum of Comparative Zoology,* 1958, *119*(7), 421–449.

Robinson, J. T. Prehominid dentition and hominid evolution. *Evolution,* 1954, *8*(4), 324–334.

Ryden, H. *God's dog.* New York: Coward, McCann, and Geoghegan, 1975.

Savage, D. E. Late Cenozoic vertebrates of the San Francisco Bay region. *University of California Publications, Bulletin of the Department of Geological Sciences,* 1951, *28*(10), 215–314.

Schaller, G. B. *The Serengeti lion: A study of predator–prey relations.* Chicago: Univ. of Chicago Press, 1972.

Schultz, J. R. A late Quaternary mammal fauna from the tar seeps of McKittrick, California. Studies on Cenozoic vertebrates of western North America. Carnegie Institution of Washington Contributions to Paleontology, 1938, *487,* 111–215.

Silver, H., & Silver, W. T. Growth and behavior of the coyote-like canid of northern New England with observations on canid hybrids. *Wildlife Monographs,* 1969, *17,* 1–41.

Simpson, G. G. Holarctic mammalian faunas and continental relationships during the Cenozoic. *Bulletin of the Geological Society of America,* 1947, *58*(2), 613–687.

Simpson, G. G. The meaning of taxonomic statements. In S. L. Washburn (Ed.), Classification and human evolution. *Viking Fund Publications in Anthropology,* 1963, *37,* 1–31.

Slaughter, B. H. *Platygonus compressus* and associated fauna from the Laubach Cave of Texas. *American Midland Naturalist,* 1966, *75*(2), 476–494.

Stock, C. Rancho La Brea: A record of Pleistocene life in California (5th ed.). *Los Angeles County Museum Science Series,* 1953, *15.*

Teleki, G. *The omnivorous chimpanzee. Scientific American,* 1973, *228*(1), 32–42.

9

Dire Wolf Systematics and Behavior

Marc Stevenson

The dire wolf of the late Pleistocene represents a third major taxon of the genus *Canis* in North America. Features that distinguish this large wolf from the modern timber wolf *Canis lupus* concern mainly the proportions of its skull and its limbs rather than discrete morphological traits. In general, the dire wolf was a more robust form than the modern timber wolf and its jaw structure was especially robust. This feature, plus the fact that it is found coeval with the extinct megafauna of the late Pleistocene, suggests that the dire wolf was ecologically dependent on the large fauna, perhaps scavenging a significant proportion of its food. Though it has been extinct for the last 8000 years, the dire wolf can be studied in order to develop models of its behavior and morphology. These models may illuminate and inform the study of early man.

HISTORY OF INVESTIGATION

In the early 1850s excavations on the banks of the Ohio River resulted in the discovery of a large Pleistocene wolf (Merriam, 1912). The species, according to Joseph Leidy, varied enough from the modern wolf form to require a distinct specific classification. First described as *Canis primaevus* and *Canis mississippiensis*, the name *Canis indianensis* became the common designation used in the literature until it was discovered that *Canis dirus* preceded it. In 1918, during an era when paleontologists approached the

179

classification of fossils somewhat differently from today, J. C. Merriam (1918b) established the genus *Aenocyon* for the dire wolf group. W. D. Matthew, however, refused to accept *Aenocyon* as a separate genus for the large Pleistocene wolf and continued to refer to these forms as *Canis*. Others also doubted the validity of a separate generic rank for the Pleistocene wolf (Merriam, 1912). Except for its relatively larger size, the dire wolf was not very different from the modern wolf of today; the difference between the wolf and the extinct wolf is not as great as that between the coyote and wolf. Hence the dire wolf should be considered in the genus *Canis*, but with subgeneric status (Allen, 1876). The common dire wolf may then be known as *Canis (Aenocyon) dirus Leidy*.

WOLF, DIRE WOLF, AND COYOTE EVOLUTION

So far as our present fossil record shows, the chief center of evolution and dispersal of the Canidae was in the northern world—in Holarctica and in the open plains country of North America (Matthew, 1930). During the Pleistocene in America the genus *Canis* is known to have been represented by the wolf, the coyote, and the dire wolf. Since the ancestry of the genus *Canis* is discussed in the preceding chapter, it will not be treated in detail here.

Very little is known concerning when the dire wolf and coyote separated from the direct line that led to the wolf and dog. In any case, the dire wolf appears to be nonexistent in early and mid-Pleistocene deposits. However, it must be noted that the absence of a particular fossil in a geological deposit of a specific Pleistocene age does not necessarily mean that it did not exist at that time or earlier. Certainly many reasonable explanations, including chance, can be offered for the absence of a particular fossil in a deposit.

If the dire wolf did in fact diverge from the main line that led to the wolf in the early Pleistocene or even later, it is reasonable to assume that the coyote separated at least as early and probably earlier. For, as noted, the chief morphological differences occurring between the wolf and the dire wolf are not nearly as great as those between the wolf and the coyote. The proposal here is not to establish absolute chronologies but rather to suggest that the coyote separated before the dire wolf. The hypothesis of a closer temporal relationship between the wolf and dire wolf is based primarily on similarity in size of prey and on the premise that the further an animal is removed anatomically from the main line of descent the further in time its separation from the main line occurred.

MORPHOLOGY OF THE WOLF, DIRE WOLF, AND COYOTE

Since the dire wolf is an extinct form, comparisons with the true wolf and the coyote are of necessity based on morphological criteria. As yet, however, it has not been possible to differentiate between the dog, coyote, wolf, and dire wolf on postcranial elements alone; size of the postcranial skeleton is of little use in distinguishing between these canids. Although a small wolf physically resembles a large coyote in postcranial skeleton, a small wolf does not assume the characteristics of a large coyote in cranial morphology. Only when cranial elements are included is it possible to distinguish between species.

The key characteristics separating *lupus* and *latrans* seem to be based on a certain intraspecific homogeneity, which is not too difficult to describe (Lawrence & Bossert, 1967). The greatest recognizable differences separating the wolf and the coyote occur in the region of the tearing teeth or carnassials—the fourth upper premolar and first lower molar. Other describable differences occur in the development of the muzzle and braincase. Wolves have a relatively massive rostrum and small braincase compared with the rest of the skull. The massiveness of the wolf skull as compared with that of the coyote presumably reflects the larger size of animals on which the wolf preys. The coyote, preying as it does on smaller animal species and birds, has a comparatively smaller rostrum in relation to the braincase and small, narrow teeth, giving it a rather long, slender appearance as compared with the wolf.

Although there are some obvious key characteristics that separate the wolf from the extinct dire wolf, the differences are not nearly as great as those occurring between the wolf and the coyote. In some dental measurements there are differences of only a few millimeters between wolf and dire wolf. Of all the cranial and dental measurements taken, the greatest differences are in the tearing or carnassial teeth and in the relative width and size of the skulls.

Compared with the wolf, the dire wolf appears relatively broader across the frontal region, palate, and zygomatic arches. The sagittal crest is extremely high and thin and is characterized by an extraordinary backward projection of the occipital bone that much exceeds that of other wolves. The lower jaw tends to be longer than in most wolves, and it is also relatively higher and thicker below the inferior carnassial teeth. The incisors and canines of the dire wolf have much the same form as those of the large recent Alaskan wolves; however, according to Merriam (1912), in most specimens the incisors are thrown considerably out of alignment by lateral crowding. The three upper premolars (P^1, P^2, and P^3) of the dire wolf tend to be relatively small and are quite similar to those of recent wolves. The upper carnassial tooth, P^4, however, is relatively larger than P^4 of the wolf. The first and second upper molars (M^1 and M^2) are practically identical with those of the modern wolf. The first three lower premolars (P_1, P_2, and P_3) also tend to be little larger than or the same size as those of the recent wolf. The fourth lower premolar (P_4) is relatively large compared with P_3 but shows exactly the same size relation to the first lower molar (M_1) as in the Alaskan wolf. Like P^4, M_1 tends to be relatively massive. The lower second and third molars, M_2 and M_3, however, are relatively small in comparison with M_1 and are to a degree similar to those of the larger modern wolves.

Most of the morphological differences existing between the wolf and the dire wolf can be accounted for by one particular evolutionary development. Wider skull proportions, the exceptional heaviness of the jaws below the lower carnassial, the extraordinary height and backward extension of the sagittal keel, and the general overall massiveness of the dire wolf skull are a direct result of the development and enlargement of the carnassial or tearing teeth. This wolf's particular primarily physiological adaptation to the larger herbivorous mammals that once roamed the Plains resulted in the enlargement of the cheek teeth. This feature in turn resulted in the heavier development of the jaw and skull as well as the heightening and lengthening of the sagittal crest, which accommodated the development of large temporal and masticatory muscles.

In many if not most respects, the dire wolf postcranially was very like the modern timber wolf except that it was somewhat larger and on the average considerably heavier than the larger wolves of today. The dire wolf may have weighed as much as 20% more than the largest Alaskan wolves. Although not much taller than the present-day wolf, the dire wolf was generally of sturdier build with a particularly long and heavy skull, strong shoulders, deep chest, and massive pelvis. Although as large as those of *Canis lupus*, the feet in proportion to the overall size of the dire wolf may have been a little shorter than those of the present-day species. However, postcranial elements should not be relied upon too heavily in distinguishing these species, for the body shape of the wolf and length of legs can be quite variable. The structure of the limbs and feet in most dire wolf fossils nonetheless suggests that this animal possessed to a lesser degree the cursorial adaptations of the canid tribe and the long-range trotting abilities of the wolf. Comparatively smaller limbs in relation to its overall weight and a massive head seem to suggest that the animal was not as well developed for running or trotting as are the timber wolves and coyotes.

The structure of the limbs also suggests that the dire wolf was relatively stronger and quicker over shorter distances than its cousin, the wolf. The stronger, heavier build of the dire wolf gives it a sturdier appearance, which presents a slight shortening of the leg and a humping of the shoulder. When all cranial and postcranial characteristics are considered, the extinct dire wolf appears to have been a near relative of the living wolf, distinguished from *Canis lupus* by its somewhat larger size and several small peculiarities in dentition and skull. (For an artist's conception of how the extinct wolf of the Pleistocene may have looked, see Figure 1.)

The dire wolf appears to have been represented in the late Pleistocene of North America by several fairly distinct and now extinct subspecies. These include

1. *Canis (Aenocyon) dirus,* a large wolf represented by a variety of forms from Rancho La Brea (Southern California) and elsewhere. It is always characterized by its immense size, massive dentition, and relatively great breadth of palate and facial region.

2. *Canis (Aenocyon) ayersi,* a large wolf found in some southern states, also characterized by large size and massive dentition. It possessed a relatively narrower and more slender facial region and a wider spacing of the premolars as compared with *Canis (Aenocyon) dirus.* It was slightly—if at all—smaller than *Canis dirus.*

3. *Canis (Aenocyon) milleri* (another large wolf from Southern California), a wolf smaller than either *Canis dirus* or *Canis ayersi.* It is characterized by a relatively low sagittal crest and a less prominent inion, a massive dentition, and closely set premolars, as is rather common in the *Canis dirus* form. The *milleri* form is described by Merriam (1912) as a distinct species having characteristics closer to those of the *Canis dirus* group than to those of the timber wolves of the *Canis occidentalis* (modern wolf) group.

4. *Canis furlongi,* sometimes referred to as *Canis occidentalis furlongi,* proposed from the Rancho La Brea region (Hay, 1927; Merriam, 1910; Schultz, 1938). This form represents a wolf considerably smaller than the smallest individuals of the *Canis (Aenocyon) dirus* types, and evidently represents a species closely related to the existing North American timber wolves. In details of tooth structure it is distinguishable from the dire wolf and the living forms of wolf by several small peculiarities. *Canis furlongi* is separated from the timber wolf on the basis of a relatively narrower palate and muzzle, and a heavier superior carnassial, yet it may have been more like the modern wolf than any other member of the extinct dire wolf group, as it almost bridges the gap between *Canis occidentalis* and the *Canis dirus* wolf. The possibility that this form may represent an interbreeding population between the extinct wolves of the *dirus* group and the

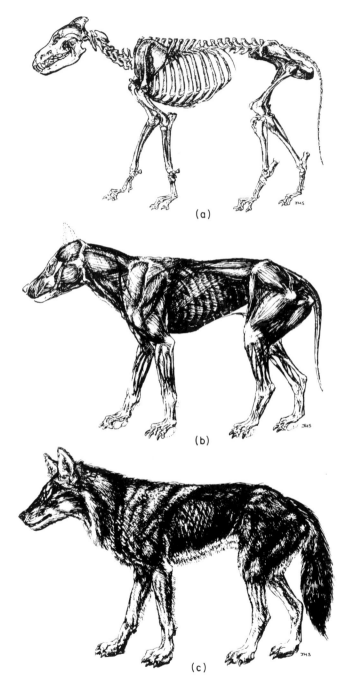

Figure 1. Artist's conception of the dire wolf: (a) skeleton, (b) muscles, (c) physical appearance. (Drawing by John Slater.)

timber wolf should be considered. Living in marginal areas may have produced wolf–dire wolf hybridizations.

PLEISTOCENE RANGES OF THE WOLF, THE DIRE WOLF, AND THE COYOTE

During the late Pleistocene in North America, the now extinct *Canis (Aenocyon) dirus* wolf was contemporaneous with the wolf and coyote. However, the representation of dire wolf in most faunal assemblages of late Wisconsin age is so far in excess of that of the wolf that *Canis lupus* may have been virtually insignificant and unable to compete. The coyote, on the other hand, appears to have been as abundant as the dire wolf in America at this time. Probably *Canis latrans* was able to exist in the same regions with the larger predatory dire wolf because it presented no formidable challenge or serious competition to *Canis dirus*. Unlike the wolf, the coyote was markedly divergent from the dire wolf in physiology and ecological preference. Coyote and dire wolf were so different that they could maintain separate identities when meetings occurred. Because the coyote and the dire wolf occupied distinctly different niches, competition between them was less intense than that between the wolf and the coyote. It may be speculated that the dire wolf had little antipathy for the coyote and regarded it more as prey or a nuisance than as a rival, whereas the wolf regards the coyote as more of a rival than as prey. Because coyotes often profit by feeding on wolf-killed carrion in harsh environments it is reasonable to assume that the late Pleistocene coyote may have been as dependent upon wolf-kills in some areas as at present.

The dire wolf was widely and abundantly distributed over much of unglaciated North America during the late Wisconsin, particularly in the latest stages of the period. Its range probably extended from the east to the west coast and from Mexico to at least as far north as the upper Mississippi Valley. Dire wolves associated with extinct fauna have also been reported in Alaska, Venezuela, and Peru.

The true wolf, however, appears to have been absent or rare in America in the Wisconsin glacial age. Except for a few occurrences in Oregon, Wisconsin, Colorado, Wyoming, and several other northern states, most of the evidence of the wolf in the late Pleistocene deposits of North America prior to 11,000 years ago seems inconclusive and questionable. The wolf during the late Pleistocene in North America does not appear to have been the significant and successful wide-ranging predator it was up until recently. However, it is probable that the timber wolf of the *Canis occidentalis* group occupied the northern portion of the continent, in the

Holarctic area and the area immediately south of the ice sheets, contemporaneously with the maximum development of the wolves of the *dirus* group in the Sonoran and southern regions of the continent (Merriam, 1918a).

Evidence that tends to confirm the extensive southern range of the dire wolf and the more limited northern range of the wolf presents itself in the Fossil Lake beds of Oregon. The large wolf of the Fossil Lake fauna seems to be exclusively of the modern timber wolf type. Although the suggestion has been raised by Elftman (1931), there has not been any concrete indication or recognition of the dire wolf species in the Fossil Lake area. Had *Canis (Aenocyon) dirus* or any other member of the group been present in the region in numbers comparable to those known elsewhere in the Pleistocene, it would presumably have left a definite trace of its presence.

The wolf appears to have inhabited much of Alaska as well as the tundra and taiga zones immediately south of the ice sheet during the maximum of the Wisconsin glaciation. The dire wolf on the other hand occupied the deciduous and boreal forest (mixed zone of transition from coniferous to deciduous forest) at the same time (Figure 2). It seems reasonable to assume that the ranges of the wolf and the dire wolf overlapped to some degree near the northern limits of the boreal forest; however, their ecological roles—major dependent predators of large game— may have been so similar that they were unable to live together in the same area. Yet it must be assumed that the niches of the two groups were distinct enough—the dire wolf adapting to relatively larger game—that their populations did not merge. The wolf and the dire wolf were most certainly engaged in competition near the limits or boundaries of their ranges. The lack of firm fossil evidence pertaining to the wolf in the interior of the Great Plains also seems to suggest that *Canis (Aenocyon) dirus* probably kept *Canis lupus* and its related subspecies in check during much of the Pleistocene.

In some cases where they met, interbreeding probably occurred. Hybridizations of the dire wolf and wolf were almost certainly as fertile as coyote–wolf hybrids of today. Their interbreeding, however, was probably only sporadic and covered a narrow zone where they overlapped.

Some late Pleistocene fossils, notably *Canis occidentalis furlongi*, may represent such dire wolf–wolf hybridizations; this species is described as bridging the gap between the timber wolf and dire wolf. The Charleston, Virginia fossil wolf jaw also demonstrates some very strong affinities with both species of wolves (Hay, 1923). In the relatively narrow zone where they would have overlapped during the Pleistocene, a number of large wolf remains demonstrate a variety of both dire wolf and wolf features.

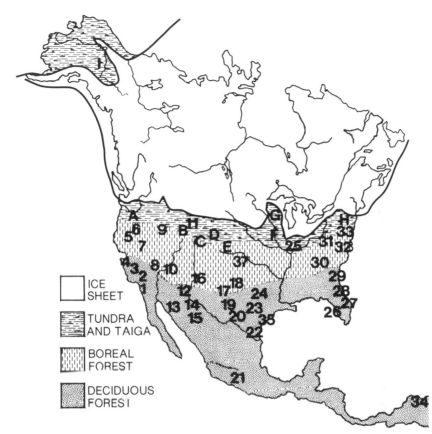

Figure 2. Wolf and dire wolf locations during the late Pleistocene (prior to 11,000 years ago). *Wolf locations:* **A,** Fossil Lake, Oreg.; **B,** Wyoming (several locations); **C,** Colorado (several locations); **D,** Nebraska (several locations); **E,** Logan County, Kans.; **F,** Alton, Ill.; **G,** Blue Mounds, Wis.; **H,** Monroe County, Pa.; **I,** Alaska. *Dire wolf locations:* **1,** Rancho La Brea, Calif.; **2,** Carpinteria, Calif.; **3,** McKittrick, Calif.; **4,** Alameda County, Calif.; **5,** Samwel Cave, Calif.; **6,** Potter Creek Cave, Calif.; **7,** Hawver Cave, Calif.; **8,** Gypsum Cave, Nev.; **9,** Jaguar Cave, Ida.; **10,** Hermit's Cave, Ariz.; **11,** Powder Mill Cave, Mont.; **12,** Whitewater Creek, Ariz.; **13,** Ventana Cave, Ariz.; **14,** Sulphur Spring, Ariz.; **15,** Murray Spring, Ariz.; **16,** Sandia Cave, N. Mex.; **17,** Blackwater Draw, N. Mex.; **18,** Briscoe County, Tex.; **19,** Midland, Tex.; **20,** Bexar County, Tex.; **21,** Tequixquiac, and Valsequillo, Mexico; **22,** Friesenhahn Cave, Tex.; **23,** Levi Site, Tex.; **24,** Lewisville, Tex.; **25,** Evansville, Ind.; **26,** Saint Petersburg, Fla.; **27,** Vero, Fla.; **28,** Melbourne, Fla.; **29,** Charleston, S. C.: **30,** Hamblen County, Tex.; **31,** Cumberland, Md.; **32,** Frankstown, Pa.; **33,** Port Kennedy Cave, Pa.; **34,** Muaco, Venezuela; **35,** Moore Pit, Tex.; **36,** Peru (not shown); **37,** Sheridan, Kans.

At some point in the Pleistocene the wolf and the dire wolf shared a common ancestor. Whether the dire wolf separated from the wolf in its fully modern form is uncertain. It is probable, however, that this ancestor, a cursorial predator, ranged over much of icefree American during an interglacial episode. When the glacial ice reemerged the large continental ice sheet most likely separated the ancestral canids, forcing them south of the glacier onto the Plains of America and north into unglaciated refugia in the interiors of Alaska and the Yukon. In isolation, the two parts of the common ancestral population made adaptions to their specific environments and the northern wolves reestablished contact with Eurasian members of their species.

EXTINCTION OF THE PLEISTOCENE DIRE WOLF

Although glacial barriers regressed far enough during interglacial episodes to permit the wolf to occupy the tundra and taiga zones immediately south of the ice sheet, only when the glaciers disappeared for the last time, approximately 10,000 years ago, did the true wolf of the north expand farther into the southern zone of North America. The fairly rapid expansion and proliferation of *Canis lupus* and its related subspecies during the early postglacial era also coincides with the disappearance and extinction of the dire wolf. By Altithermal times, the representation of dire wolf begins to vary inversely with the wolf. The *dirus* wolves show a gradual decrease in numbers whereas the wolf rapidly expanded its numbers and territory.

Why was the dire wolf not able to survive and what led to its extinction? The extinction of the Pleistocene wolf may have been due to its inability to adapt to the demise of the larger herbivores. This predator seems to have been unable to adjust adequately to some great change that came about with too great speed and was perhaps marked by a decrease in food. Early post-Pleistocene climatic fluctuations which directly affected biotic communities through increasing extremes in rainfall and temperature probably have to be considered the primary agent reasponsible for the extinction of the Pleistocene megafauna. Increasing evidence (Yapp & Epstein, 1977) suggests that rainfall and temperature during the late Pleistocene were seasonally more equable and stable than present. The occurrence of "disharmonious" faunas in many Pleistocene deposits in North America also suggests a stable climate, with less marked seasonality in rainfall and temperature. The stable climates and environments of the Pleistocene appear to have favored a biological strategy in some mammals, which selected for an increase in body size and perhaps a decrease in reproductive activity. When conditions became more unstable at the end of the Pleistocene the megafauna and their predators appear to have been too far off on this strategy to adjust back to a strategy that called for a decrease in body size and faster reproduction.

Canis (Aenocyon) dirus faced extinction because the larger herbivorous mammals and edentates—such as the mammoth, mastodon, camel, horse, giant sloth, peccary, and giant bison—with which it has been associated in geological deposits became extinct. The dire wolf it seems could not adapt to, or survive upon, the smaller ungulates such as the wapiti, moose, caribou, musk-ox, and the smaller forms of bison, which did not experience evolutionary extinction and which were replacing many of the larger extinct mammals that once roamed much of unglaciated America. These smaller mammals may have survived the rigorous, unstable, and fluctuating climates of the early post-Pleistocene because they may have reproduced faster than the larger herbivores. Those surviving also may have been relatively more fit, for they, unlike the extinct terrestrial giants of the Plains, may have been less sedentary and more accustomed to foraging and traveling greater distances in search of food. Under the fluctuating climates of the late and early post-Pleistocene, well-established feeding grounds and favorable areas of vegetation may have been upset, yet ability of the smaller mammals to reproduce faster and travel farther may have saved them and their major predator, the wolf, from extinction.

The wolf also seems to have possessed the hunting ability and social organization to survive in a changing and unstable environment. It is postulated that the wolf's evolution in a relatively more severe environment brought about features in his social organization, such as flexibility and spatial–temporal territories, which enabled it to survive. The dire wolf could not adapt to, or did not have enough time to adapt to, the new predominant megafauna, and we may surmise that it lacked some important features in its social organization, hunting habits, and life patterns to deal with the new situation. In other words, the dire wolf seems to have possessed a less adaptive social organization during the early post-Pleistocene.

A MODEL FOR DIRE WOLF SOCIAL ORGANIZATION

The problems faced in discussing the social habits of the dire wolf are similar to the problems encountered when discussing the social behavior and abilities of early man. When we make behavioral inferences we are treading on conjectural ground. It is difficult to make the transition from bones to behavior especially in dealing with highly intelligent forms, and especially when contrasting the behavior of animals having very similar physiologies. Therefore, the pattern developed to illustrate the dire wolf's social organization should be considered only as one hypothetical model.

One aspect of this model assumes that *Canis (Aenocyon) dirus* seems to have been so well adjusted biologically to the large herbivores that energy relationships and demands alone may have made it impossible for it to adjust wholly to and subsist upon the smaller and swifter megafauna that were gradually occupying the regions vacated by the larger terrestrial mammals. Lacking the highly developed trotting and running abilities of the wolf, the dire wolf may well have hunted in large packs, preying on the more burdened and slower moving animals with which he has been associated. The great numbers of these individuals found at Rancho La Brea suggest that the wolves of this species sometimes associated themselves in groups of considerable size to kill isolated ungulates and edentates. Although still possessing the slashing bite of the wolf, *Canis (Aenocyon) dirus* seems to have had a propensity toward bone smashing. This habit, however, had not modified the form of the teeth, except in a relatively larger dental development. It may be postulated that the dire wolf, being somewhat less mobile and far-ranging than the modern wolf, resorted to carrion feeding (the eating of dead, putrefying flesh) during the late Pleistocene when its prey began to decline. Carrion eating is considered, to some extent, a degenerate adaptation and a sort of parasitism, and the species that take to it do not tend to survive in competition with self-supporting carnivores (Matthew, 1930).

Although the dire wolf and the wolf were large predators with similar lifeways—both were adapted to large prey species—*Canis (Aenocyon) dirus* possibly did not have the social behavior and social organization to cope with the unstable late Pleistocene environment, or to adapt to the relatively new forms of fauna appearing on the Plains. Even if the dire wolf had the necessary hunting skills, behavior, and social features needed, its survival would have been difficult for the process of predation does not come about simply as a result of random contacts between predator and prey. Adaptations between predator and prey (Pimlott, 1967) are complicated processes that probably developed in relatively stable environments over long periods of time.

Wolves are highly social and gregarious animals; the dire wolf, on the other hand, may have had lesser capacity for social behavior. If we make the ethological assumption that even complex animal behavior reflects at least some component of genotype, then we must expect social systems to have evolved just as structure and physiology have evolved (Barash, 1974). As products of evolution, social systems should thus be adaptive; that is, they should be peculiarly adjusted to each environment in which they occur so as to confer maximum reproductive success (Barash, 1974).

With its range restricted to the northern refugia of the Yukon and Alaska, and the tundra and taiga zones immediately south of the ice sheet during the Pleistocene, the wolf undoubtedly faced a much more severe environment than the more southerly distributed dire wolf. This situation would have favored increased sociality, colonial organization, late dispersal, and social tolerance. Age of dispersal, sexual maturation, and growth rate would have been delayed by severe environments simply because proportionately greater maturity might have been required for success in such environments. The wolf's social evolution in the northern world would also have involved a corresponding decrease in aggressive behavior toward members of its own species, and an increase in social tolerance to accommodate the animal's need to disperse at an advanced age. The progressive increase in sociality among animals experiencing progressively more severe climates may be due to the increasing necessity to inhibit the dispersal of undersized animals (Barash, 1974).

The evolution of the dire wolf's social system in the less severe environments of the south did not require the evolved social organization of the wolf (Figure 3). The dire wolf's evolution in the less severe, more stable environments of Pleistocene America also meant that it may have developed more stable relationships with its prey. Clearly this situation would have been less adaptive during the early post-Pleistocene. Had the dire wolf evolved a social system comparable to that of the wolf, his chances of survival during early postglacial episodes may have been increased to the point where he could have dealt with the difficult conditions and unstable environments of the late Pleistocene.

MODELS FOR EARLY HOMINID EVOLUTION

As noted elsewhere in this volume, many ecological and behavioral factors promote the study of North American canids as analogous to early hominids. Most of these factors stem from the hypothesis that hunting was the major factor in man's cultural evolution. It is argued that the wolf, dire wolf, and coyote may have evolved in similar directions to the Plio-Pleistocene hominids, and that the study of these canids can be useful in developing models for early man. Behavioral and physiological differences existing in the North American members of the genus *Canis* provide an interesting framework that can be compared with the pattern of variation seen in several groups of fossil man.

Two models are elaborated here. The first is developed irrespective of existing hominid fossil data. The second model, which is similar to one model proposed by Hall (Chapter 8), attempts to account for the variation

Figure 3. Dire wolves—a fanciful reconstruction of a meal-time disagreement. (Drawing by John Slater.)

and speciation in wolves and hominids by comparing their patterns of variation.

First, let me suggest that some species of early man may have evolved and receded like the dire wolf, serving as temporary evolutionary experiments. These experimental groups, having radiated from the major stock, evolved rapidly and diversified into new niches that proved viable for some time. Although some hybridization occurred, the niches of the experimental and major groups were distinct enough that the groups did not merge. Interbreeding was not significant enough to make anatomical impacts upon either the experimental or major group. Behavioral mechanisms and ecological preferences kept the experimental forms evolving separately in different directions. Like the dire wolf these evolutionary experiments developed relatively innovative and stable relationships with their environments—perhaps too stable, for when conditions changed neither species had developed the necessary social, behavioral, and physiological tools. The fate of these experimental forms was extinction.

The second model derives from increasing fossil evidence, which suggests that at least three morphologically distinct groups of early man existed in Africa for several million years. The simple progression from *Australopithecus* to *Homo* no longer seems valid. Although some have argued that all early hominid fossils represent geographical or regional variants of the same species, there seems to be fairly good evidence for three distinct species of early man. These include: *(a)* a robust form, *(b)* a gracile form, and *(c)* an intermediate form, tentatively referred to as *Homo* sp. or *Homo habilis,* which more closely resembles *Homo sapiens* than either the gracile or robust species.

The existence of several early hominids and the pattern of variation occurring between them can be compared with the relationships existing between the wolf, dire wolf, and coyote. Like the Plio-Pleistocene of Africa there existed in the Pleistocene of North America ecological space for three different but closely related social predators. The dire wolf and coyote were generally adapted to the southern deciduous forest and semiopen brush environments of America, whereas the wolf was restricted to the tundra and northern timbered regions of the continent. Although some geographical, anatomical, and behavioral overlap occurs within these three groups of canids, their behavior, habits, and general overall morphology were sufficiently distinct that each group was able to maintain itself. Each of the three species avoided competition with the others by seeking different prey and, with the exception of the dire wolf and coyote, by inhabiting different environments. Each canid was to some degree anatomically and behaviorally suited to its prey and niche (Fox, 1975).

The dire wolf and coyote were able to coexist contemporaneously in the same environment because they were morphologically and ecologically divergent enough to maintain their separate identities. Their coexistence during the Pleistocene was also facilitated by an abundance of many different prey species ranging greatly in size. Their markedly divergent physiologies enabled them to inhabit different niches in the same habitat—the dire wolf adapting to larger herbivorous prey and the coyote adjusting to much smaller prey. It tentatively is postulated that both species of early near-man may have been able to coexist in the same environment for the same reasons that the dire wolf and the coyote were able to coexist.

Notwithstanding the fact that ecological preferences have a morphological basis, the separation of the dire wolf from the wolf, and the coyote from the wolf, was at first enforced predominantly by behavioral features. With the dire wolf and coyote being closer to the wolf anatomically and ecologically than to each other, habitat preferences based on behavioral mechanisms such as mutual antipathy and different hunting methods, rather than morphology, kept the dire wolf and coyote evolving in separate directions from the wolf. I suggest that, as with these canid counterparts, ecological preferences and behavioral features would have at first kept the gracile and robust hominid species evolving apart from *Homo*. Like the wolf, our ancestor may have been anatomically and ecologically too similar to either the gracile or robust members of its group to occur in the same habitat. Mutual antipathy and methods of hunting alone were not strong enough or divergent enough for two physiologically similar species to inhabit the same environment.

Perhaps like the coyote, the gracile hominid was an efficient individual hunter of small prey who occasionally formed in larger parties to obtain food by cooperative hunting techniques. Morphology, however, indicates that the coyote and the gracile form were incapable of systematically exploiting larger game, without cooperative hunting methods and advanced social organization. Because the abundance and size of prey directly influences the method of hunting, the smaller animals that these two species preyed on usually did not require cooperative hunting methods.

It is postulated that meat contributed to a significant portion of the diets of the three early hominids but the size of prey and method of hunting differed. Following the wolf–dire wolf–coyote model, it is proposed that the robust and intermediate hominid forms ate larger and heavier game, and that the robust form relied more on its physical abilities in the hunt than on cooperative hunting methods. Our ancestor, however, did not adapt to larger prey strictly in an anatomical direction; rather, it was able to transcend the prey–predator body size relationships by developing cooperative hunting methods and the social behavior and organization that goes along with them. The gracile hominid, on the other hand, adjusted to smaller prey that did not require size or the use of cooperative hunting techniques. In summary, each hominid, as each canid, was to differing degrees anatomically and behaviorally suited to its prey and niche (Figure 4).

CONCLUDING REMARKS

In some detail, I have elaborated a reconstruction of the evolution, behavior, habits, and extinction of the dire wolf. Because very little information regarding the dire wolf exists in literature, several models consistent with the data were developed so that comparisons with known canid morphology, behavior, habits, and patterns could be made. Together with the known patterns and ranges of variation in the coyote and wolf, these schemes were worked into a practical model for the evolution and behavior of three early hominids. It has also been argued that some

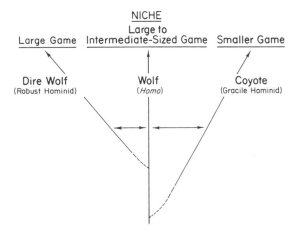

Figure 4. A framework of hominid evolution devised from *Canis*.

canid and hominid forms experienced extinction because they adapted primarily in physiological or biological directions.

I suggest that, as in late Pleistocene North America, there existed in Plio-Pleistocene Africa ecological space for three highly intelligent predators with varying degrees of sociability and adaptability. A subtle, reciprocal anatomical and behavioral relationship occurred between each canid and hominid predator and its prey. However, it is postulated that our ancestor survived because it adapted primarily in a sociobehavioral rather than a biological direction. When conditions changed, the gracile and robust hominid forms, lacking an adaptive and developed social organization and behavior, faced evolutionary extinction.

It is also of interest to note that the wolf has been recorded on every inhabitable continent except Australia and Africa. The absence of the wolf in Australia can be explained by this continent's relative isolation from the mainland of Eurasia; however the absence of the wolf from the Plio-Pleistocene deposits of Africa is not so easily explained. It is suggested that the wolf stock was not able to penetrate successfully into the African continent because there was no ecological space for another highly organized social hunter of large game.

REFERENCES

Allen, J. A. Description of some remains of an extinct wolf and extinct species of deer from the lead region of the Upper Mississippi. *American Journal of Science*, 1896, *2*, 47–51.
Barash, D. P. The evolution of marmot societies: A general theory. *Science*, 1974, *185*, 415–420.

Bueler, L. E. *Wild dogs of the world.* New York: Stein and Day, 1973.

Elftman, H. O. Pleistocene mammals of Fossil Lake, Oregon. *American Museum Novitates,* 1931, *481,* 1–21.

Fox, M. W. Evolution of social behavior in Canidae. In M. W. Fox (Ed.), *The wild canids.* Princeton, New Jersey: Van Nostrand-Reinhold, 1975.

Hay, O. P. The Pleistocene of North America and its vertebrate animals from the states east of the Mississippi River and from the Canadian provinces east of longitude 95°. *Carnegie Institution of Washington,* 1923.

Hay, O. P. Pleistocene of North America and its vertebrated animals, west. *Carnegie Institution of Washington,* 1927.

Lawrence, B., & Bossert, W. H. Multiple character analysis of *Canis. American Zoologist,* 1967, *7,* 223–232.

Matthew, W. D. The phylogeny of dogs. *Journal of Mammalogy,* 1930, *11*(2), 117–138.

Merriam, J. C. New mammalia from Rancho La Brea. University of California Publications Bulletin, Department of Geology, 1910, *5*(25), 391–395.

Merriam, J. C. The fauna of Rancho La Brea, Part II. *Memoirs University of California,* 1912, *1*(2), 217–272.

Merriam, J. C. Evidence of mammalian paleontology relating to the age of Lake Lahontian. *University of California Publications Bulletin,* Department of Geology, 1918, *10*(25), 517–521. (a)

Merriam, J. C. Note on the systematic position of wolves of the *Canis dirus* group. *University of California Publications Bulletin,* Department of Geology, 1918, *10*(27), 531–533. (b)

Pimlott, D. H. Wolf predation and ungulate populations. *American Zoologist,* 1967, *7,* 267–278.

Rand, A. L. The Ice Age and animal speciation in North America. *Arctic,* 1954, *7,* 31–35.

Schultz, J. R. A late Cenozoic vertebrate fauna from the Coso Mountains, Inyo County, California. *Studies on Cenezoic Vertebrates of Western North America,* Carnegie Insituation of Washington, Contributions to Paleontology, 1938, *487,* 75–109.

Yapp, C. J., & Epstein, S. Climatic implications of D-H ratios of meteoric water over North America as inferred by ancient wood cellulose C-H hydrogen. *Earth and Planetary Science Letters,* 1977, *34,* 333–350.

CONCLUSION

Wolf and Human

Anthropologists ponder the means by which cultures ensure that activities, ideologies, and personalities are sufficiently integrated to safeguard their own continuance. Few doubt that most cultural systems are adequately, if not optimally, integrated to meet the basic physical and psychological needs of the individuals that comprise the cultural group. But *how* do cultural systems achieve this? What makes them tick?

Basically, this nuts-and-bolts question is unanswerable unless we realize that there exist many levels of explanation for human behavior. As scientists we cannot expect to discover the "final truth" but can only specify more and less useful frames of explanation.

Anthropologists often disagree on how to explain or account for variability in cultural behaviors. Do you ask the people who live in a cultural system to tell you? Do you assume that their behavior is purely a matter of learning? Should you take an economic determinist stand and claim that a struggle for the society's resources determines whatever cultural behaviors, including beliefs, are practiced? Are Kurt Vonnegut's "Tralfamadorians" to blame—some outside stand-in for the mysterious variable that is as yet undiscovered?

Even though anthropologists disagree greatly, they (as a group) agree that cultural behavior must be studied apart from the biological group that practices it. In general they are wary of attempts to explain cultural behavior in terms of the genetic structure of particular persons, or in terms of biochemical processes that occur within those persons. Indeed, to call an anthropologist a reductionist—meaning one who reduces complex social behaviors to neurological or biochemical processes—is the worst kind of insult. Unfortunately, this term, like any other kind of invective, tends to be applied liberally and some anthropologists apply it to any theorist with whom they disagree. This is especially unfortunate if a serious and novel approach to the problem of human behavior does not receive the attention it deserves because it has been capriciously labeled as reductionist.

Historically the most obvious reason for this strong bias against cultural explanations in terms of the biological infrastructure was the necessity to free the discipline from racism. In the nineteenth century, attempts were made to explain differences in human cultures by (supposedly) significant biological differences among human populations. In recent years it has been demonstrated overwhelmingly that the biological differences that do exist among the multitude of local populations of *Homo sapiens* are unrelatable to psychological or cultural differences.

In addition to stipulating that culture must be interpreted as a phenomenon in and of itself, separate from the biological traits of the persons who practice it, anthropologists tend to limit culture to one species, *Homo sapiens*. We are questioning this belief throughout this book. Human culture as a total phenomenon is different from culture as practiced by other species, but it is by no means clear that humans alone bear culture.

It is our contention that in looking for evidence of culture we have paid too much attention to the capacity of humans to make statements about their culture. An individual's explanation of his behavior and the practices of his culture is useful only at certain levels of explanation. Verbalizations are interesting and significant phenomena in their own right but they do not necessarily help an outside observer understand how a social system works at all levels of analysis. This does not mean that verbalized explanations are functionless. Consider our day-to-day conversation about politics in our national capital. Most of us recognize that our chitchat does not provide a good or reasonable explanation of political behavior. The ability to expound on a variety of topics is a social skill, and serves social functions; this ability is not a necessary part of culture, in a general sense.

We noted in the Introduction to Part I that two criteria (the existence of social systems that vary according to cultural criteria, and the use of symbolic communication) are most critical in recognizing or diagnosing the presence of culture in any animal population. Let us review each of the chapters specifically to determine the support that each provides for this point of view.

BEHAVIOR AND CULTURE

In Chapter 1, Man, Wolf, and Dog, Fox reminds us that, like other animals, we are part of an evolutionary sequence. He points out similarities between the processes of domestication of the dog and culturation in humans. In Chapter 2, Variability in the Wolf, a Group Hunter, Sullivan is concerned with the behavioral variability shown by individual wolves and by wolf packs, and looks at four aspects of wolf adaptation: group hunting, aggressive behavior, reproduction, and individuality.

In Chapter 3, Natural History of the Coyote, McMahan shows the coyote to be an intelligent animal adaptable enough to reduce social involvement with its own kind to fare better in a life-and-death struggle with humanity. It is possible to describe what individual coyotes do to aid their own survival in the face of heavy human predation. At a more comprehensive level of analysis we can demonstrate the effect of individual behaviors on the total coyote population. For example, the tendency to separate while still young reduces the level of visibility and ensures that no more than one coyote is caught at a particular spot. This practice tends to increase the total coyote population as each female who can find a mate rears a litter. In contrast, when coyotes live in packs, usually only one female produces young.

From the observer's perspective we can analyze the coyote's natural history in terms of individual and group reactions to human predation. We can also judge the systemic effects of coyote culture and appraise it as a total survival strategy. The fact that individual coyotes may not conceptualize a life-strategy in the same way that we do should not detract from our judgment as to whether or not coyotes have culture, any more than we should be bound by a person's concepts in our analysis of a human culture.

Similarly, as Sharp points out in Chapter 4, Comparative Ethnology of Wolf and Chipewyan, Chipewyan members of cultural subgroups act out their roles in what appears to an outsider to be an optimal strategy for ensuring that some, and possibly many, Chipewyan will secure enough caribou to continue their biological population and maintain their culture. Yet the functioning of the total system is not dependent on conscious awareness of the functioning. This assertion is reinforced by the observation that many native cultures were destroyed by the adoption of seemingly minor Euro-American culture traits, which eventually undermined the ideological or resource base of the culture. Had the practitioners of the culture (the culture bearers) perceived the effect of adopting new behaviors they would not have (willingly) participated in its destruction.

The test of whether a behavioral pattern is or is not cultural cannot be made on the basis of whether participants in the culture perceive its complexities. If this were so, no human society could be considered to have culture. This is as true for the twentieth-century United States and Soviet Union as it is for the nineteenth-century Chipewyan. Try as they might, the communist countries have not yet been able to understand their own system well enough to control it. Nor have economists in capitalistic societies of the West been able to predict, let alone control, the human behavior that in its intricate interaction makes up the socioeconomic system.

Indeed, cultures such as those of the prehistoric and early historic Chipewyan and the modern timber wolf—in which selection over many generations has tuned the system to harmonize with the resource base— seem to function best from an objective viewpoint. Consciousness of the total system and intention to produce a working society are not good indicators of whether a system is cultural or whether it is efficient. (And what is efficiency but a culture-bound value judgment?)

COGNITION AND COMMUNICATION

What about cognition and communication, the subjects of Part II? Peters shows, in Chapter 5, that the phenomenon we call cognitive mapping—the ability to use insight and strategy in inventing new routes across terrain—exists in humans and in other animals too. We should expect this trait to be selected for in wide-ranging predators such as early hominids and modern wolves. Holloway (1969) believes that the kind of cognitive processes involved in mapping provide a cognitive evolutionary model of language structure. But in the comparison of human and nonhuman communication systems it is best not to use language; we suggested earlier that language has more significance for human behavior in what it has done to the process of human thought than in how it functions to transfer information, which (strictly speaking) denotes communication.

Wolf vocalization is rich in variability and content. As Harrington and Mech show in Chapter 6, we have only begun to interpret it. Vocalizations are not merely responses to stimuli. They are patterned to serve as an integral aspect of the social system and a tool of wolf culture.

Peters and Mech, in Chapter 7, show that scent-marking is used by wolves not simply to mark territory but to transfer information. Analogous forms of communication occur among human groups. In the Western Desert of Australia, aborigines read and interpret the human footprints they cross in the sand. From the tracks they can determine how many persons made up a party and what kind of activity took place—a party in transit, a short stop to eat a meal, or an overnight camping visit. When a vengeance party set out on the desert the people wore feathers on their feet to disguise their footprints. In this example, recorded by Australian anthropologist Robert Tonkinson (personal communication) the marking was performed incidentally rather than deliberately. It is *interpreted* by those to whom the information is useful. The people know that enemies as well as friends might read the inevitable footprints but only when deceit is desirable does there exist conscious concern about the communicative function.

Similarly, most of us leave traces of our activities without intending to. Members of our families interpret these traces if it is useful to—for example, "She took her briefcase so she must be at the library rather than at the grocery store." The point of these examples is once again to dispel the bugaboo of conscious, recorded intent as solely indicative of the communicative act.

All animals are constant communicators in the sense that they are at all times, by their behavior, sending out information to the world outside themselves concerning their activities and internal states. The issue of cognition enters the picture when we consider the interpretation of the communicative acts. Can the perceiving animal determine not only the presence of another animal but what it is doing and what its next-planned activity is? Thus the footprint becomes more than a mark on sand. It is an indication of another individual's activities that may be useful to the observer. The wolf's urine is more than waste liquid; it is an indicator of the activities of other individuals the observer may wish to interact with or avoid.

Some anthropologists have argued that only human communication includes symbols as well as signs, that is, has arbitrary meanings as well as meanings automatically perceived. (To contrast symbol and sign let us consider the sentence, "I feel pain." This sentence may be considered symbolic; a sharp cry emitted when an individual is struck may be considered a sign.) The complexity that we are now beginning to recognize in scent-marking and in wolf vocalizations suggests that, though they include signs, they also include symbols. The total patterning and the use made of the signs within the social system are symbolic. The distinction between symbol and sign does not serve to measure whether a system is cultural. Indeed, the continuation of its use seems intended only to increase humanity's isolation from other members of the animal kingdom, rather than to help us understand the forces that have shaped us.

PALEOBIOLOGY

Basic to Part III, Paleobiology, is the argument that the classification of early hominids suffers because the concept of species has been applied too simplistically. The species concept has a long but unstable history in biology. Several hundred years ago, categories of animal groups were defined morphologically—according to outstanding observable attributes—because the model explaining animal diversity was, simply, creation. Uniformity and stability of populations were expected traits; diversity in animal members of any particular group was considered accidental, the sort of thing one must accept in an all-too-imperfect world. Still, the seventeenth-century naturalist John Ray suggested that the proper definition of *species* is one based on reproductive commonality (Greene, 1959).

Ray's point was simple but important. He said that in many groups there are two different forms, male and female, with strikingly different physical traits. Yet we have to consider brother and sister as in the same species regardless of how different they look. Other polymorphisms exist, besides the basic sex dimorphism, and the species concept should be defined to reflect these realities.

Ray's sensible suggestion was largely overlooked until Darwin's time, when, after the publication of *Origin of Species* in 1859, the evolutionary model began to replace the creationist model (at least in scientific circles). The Darwinian model presumed variability as a natural, expected (but still unexplained) element in biological reproduction. Indeed, natural selection and its concomitant result, evolutionary change, *require* variation with which to fashion new species from old. The origin of a new species began to be understood as a process in which reproductive commonality within a large group of animals became fragmented; a new species was traced back to a reproductive gap. Given enough time, a break in the simple exchange of mates between two groups results not only in new species but in new orders of animals. Indeed, it lies at the heart of the Darwinian explanation for the tremendous variability existing in the animal community through time and space.

From a theoretical standpoint, then, the evolutionary model *requires* that a species be defined as a group with reproductive integrity. As the broadest reproductive unit within which genetic material can be exchanged, the species is the unit within which the forces of evolution—mutation, natural selection, and the chance loss of genes—are felt and expressed.

However logical and felicitous the concept of the species as a tight reproductive unit may be, field naturalists have not always found it useful in actually classifying species. Once again, it is wise to recall that a "law of nature" is not a law that forces the creatures or elements of the natural world to conform to its dictates, but a principle that guides scientific perception of the natural world. In the example at hand, field naturalists and taxonomists have found that the species concept is useful most of the time but that in some cases empirical data do not conform exactly to the model—for example, in the taxonomic controversy about the southern red wolf, and the New England coy–wolf. The problem is even more acute, however, in classifying extinct animals, because it is empirically impossible to test classificatory hypotheses.

In its most ludicrous form the issue can be satirized as a kind of paleovoyeurism. We can caricature one group of paleoanthropologists (the splitters) as maintaining hotly that the separate groups of aus-tralopithecines and the early *Homo* forms would not *consider* mating outside their own ranks. Meanwhile the opposing camp (the lumpers) maintains that the australopithecines chose mates democratically, with genes flowing freely from one population to another. When considering one's own (human) ancestors there is a tendency to reconstruct their morality and sexual systems to include a flavor of one's own political and personal morality. Thus, recognition of many separate species appears to the splitter to be simply preferential mating with a biological base, but it looks like apartheid to the lumpers. At the other extreme, recognition of only one australopithecine species appears a chaotic and improbable biological "democracy" to the splitters, but it is simple random mating for the lumpers.

To the satirist, extreme concern with the mating system of long-extinct hominids appears to be paleovoyeurism; but we have seen that from the viewpoint of theoretical biology the concept of the species as a sexually closed reproductive unit is sound. It merely requires some "softening" to bring it into line with empirical reality. And what better animal model than that of North American canids, whose basic demographic and ecological adaptation has so much in common with that of the early hominids?

Hence, in Chapter 8, Hall suggests that the genus—the category just above that of the species within the taxonomic hierarchy—be considered a group of species with a similar adaptive zone and a close biological relationship. Species included within the same genus must be recognized, paleontologically, by morphological criteria, and it may be assumed that slightly different behavior characterizes each recognized species. As one aspect of each species' unique morphological and behavioral adaptation, it is assumed that species would tend to restrict mate choices to their own population. Hence, evolutionary novelties that occur in one group would not spread outside the population.

On occasion, however (particularly in wide-ranging and highly intelligent species who depend on learned behavior), reproductive barriers break down. Usually, breakdown does not bring an end to the speciation process but is temporary. The key to each particular situation turns on the ecological forces that supported the origin of the two species in the first place. So long as the most efficient means of exploiting a given set of environments results from two (or more) animal species applying slightly varied behavioral strategies, occasional cross-mating will not have a long-term effect. This is the situation that prevails among the modern North American canids, and we may assume it also held for the Pliocene hominids.

Anthropologists have concerned themselves for too long with plotting rigid (and untestable) family trees for themselves, and have overlooked biological regularities in the diverse group of early hominids. Both Hall's chapter and the following chapter by Stevenson (Chapter 9) represent exploratory attempts to present this viewpoint. Hall uses coyote–wolf data to interpret the australopitchecine material, and Stevenson, conversely, uses hominid as well as canid data to interpret the behavior and morphology of the dire wolf.

THE PATH AHEAD

We began this book with a simple idea we wished to explore. We thought the behavior, ecology, and social organization of the wolf would be a useful source of models for hominid behavior. When one of us (Hall) began to investigate coyotes we realized that as conspecifics of the wolf they too had to be included as sources in our search for models. Since we had no firsthand data, and the astonishingly small amount of published data did not deal with the issues crucial to anthropologists, we decided to try to solve the problem by soliciting manuscripts from potential contributors of other disciplines. Not until the manuscripts arrived did we fully realize the theoretical problems raised by our "simple" problem.

Our intent was to be interdisciplinary but we soon realized that the disciplines concerned have not been communicating. The sociological disciplines dealing with human social behavior are remarkably ignorant of the premises of the biological disciplines so heavily influenced by various forms of behaviorism. Those sociologically inclined formulate their theories on the most outrageous biological beliefs, often ignorant of physiology, genetics, and learning processes. Models tend to be formulated for a mythical beast, biological man, that has little resemblance to the organism we call *Homo sapiens*.

Our colleagues in the biological sciences are our equals in ignorance. They suffer from a twofold source of error when dealing with human social behavior. Their conception of man as a cultural being is based upon humanistic assertions based in our culture's symbolic systems. Ironically, the blind spots of the fields are complementary. The biologist has a sophisticated understanding of the organism but little understanding of the organism's behavior as a social animal. The sociologist has a sophisticated understanding of the organism's social behavior but little understanding of the organism itself.

It is obvious that both fields share a basic misunderstanding of humanity that is derived from the cultural matrix of which they are a part (i.e., the nature–culture opposition so fascinating to the social anthropologist). Parts of the unrecognized conceptual material can fall *beneath* the specific training and research of the scientist but *beyond* the bounds of that training and research, so that it conditions and constrains the very scientific disciplines themselves. If we are to advance our understanding of our subjects we must face this issue and break away from the mythology of our own culture.

A clear illustration of these constraints can be obtained by contrasting the chapters of Fox and Sharp. Both deal comparatively with human and animal and both seek a fuller understanding of both species in their effort. Fox (Chapter 1) writes from a biological perspective that is based on individual behavior. He seeks understanding of humans through comparison of individual behavior in different species. Insight into social behavior is gained through generalizations of this behavior in animal species and analogs and context variations in human species. Always the assumptions are based on individual behavior and the argument is from biology–psychology to social behavior.

Sharp (Chapter 4) works from the opposite perspective. He starts with social behavior and sociological models derived from human societies and seeks insights by applying them to wolf and human as common predators of a single region. The theoretical perspective here is always sociological; the analysis is of the behavior and relationships between groups, not between individuals. Lack of material prevents the analysis from continuing into comparative social structure (this lack of material desperately needs correction!) but always it is the social group that is analyzed, not the individual.

A comparison of the methodologies and theoretical assumptions of the two chapters is perhaps more significant than the success or failure of the separate attempts. Few people today challenge the idea that much is to be learned about human behavior from comparison with animal behavior. But what of the opposite proposition? Even if our valuation of our own nature precludes a serious attempt along this line, the fact remains that sociological analysis, the analysis of social behavior as a social process, has been restricted to human societies. Even the most sophisticated analyses of animal behavior are, in one sense, reductionistic and have been based on individual behavior. But it is not possible to explain social behavior in terms of individual behavior or even from the perspective of a pair or of a single social group. We all agree that culture is a product of evolution. We all agree that numerous animal species are social—that is, that the individuals must belong to a social group to gain their subsistence. But where is the field of animal sociology? That it does not exist can only be due to our culture's valuation of our uniqueness, a set of values that has resulted in the *a priori* assumption that animal behavior cannot be explained in sociological terms because the social behavior that sociology deals with does not, by definition, exist among animals. We must learn to talk and learn across discipline lines.

In this sense Sharp's analysis is crucial to our book. If the application of theoretical material derived from social anthropology can be applied to a wolf social system without abrogating any basic theoretical postulates, it indicates that we have overlooked an aspect of the behavior of social animals without which we cannot hope to arrive at a degree of sophistication in explaining animal social behavior. Also, by postulating and defending the existence of phenomena analyzable in these terms, our argument about the crucial role of social hunting in human evolution is supported.

The judgment as to success or failure must be left to the reader but we feel that sociologists and anthropologists will see occasions in the superb work of Mech, Peters, and Harrington where reference to sociological models would ease the burden of interpretation.

Familiarity with human social systems causes the investigator to be aware of many variations in practice, belief, and language even in small social groups. These variations, like variations in a gene pool, provide competing ways of dealing with social life and the cultural condition at many levels. It is tempting to generalize, after due consideration of the canid material here—for both living animal and fossil—and postulate that such variations are part of all social systems, human and animal. Obviously our concepts of species, gene pool, and population are not as clear and compatible as we would like to think. However we classify the canids of North America, they show the variability we have described. The wolf is a social hunter of large game, the coyote a hunter of small game with a social system based upon pair-bonding. The dog, a direct exploiter of man, is not even a social animal in its domesticated condition. (Obviously social dependence is not genetic in origin in canids! So can we assume it is in primates?)

But variation exists within canid species as well as between them. The state of the wolf literature is comparable to the primate literature at the time DeVore and Washburn (1961) articulated their dominance model of baboon social organization. That this model has been superceded does not invalidate the point that no comparable model exists for wolf social organization. Wolves in Alaska's McKinley Park do not have the same social organization as wolves on Isle Royale; Isle Royale wolves differ greatly from Minnesota wolves, which in turn differ from wolves in Algonquin Park, Canada. Captive wolves in Chicago behave unlike any of their wild brethren. No model has been forthcoming because wolves are cultural animals with a learned—and hence variable—social organization. Their social organization in different places, even under similar ecological conditions, is neither identical nor entirely predictable.

Recognizing that wolves are cultural animals raises a great many questions for an anthropologist. How useful will cultural and sociological models be? Is reciprocity operative? Kinship? Incest? Exogamy? We cannot answer these questions or even pose all of them. We can only refer our colleagues who have relevant data to the anthropological literature and to the anthropologists themselves. As we begin to realize that the difference between ourselves and animals is not what our culture tells us it is, we must also realize that many disciplines must be utilized if we are to gain any real understanding of ourselves and our coinhabitants of this planet.

REFERENCES

Greene, John C. The death of Adam. Ames, Iowa: Iowa State Univ. Press, 1959.

Holloway, R. J., Jr. Culture: A *human* domain. *Current Anthropology*, 1969, *10*, 395–412.

Washburn, S. L., & DeVore, I. Social behavior of baboons and early man. In S. L. Washburn (Ed.), *Social life of early man*. Viking Fund Publications in Anthropology, No. 31. New York: Wenner Gren Foundation, 1961. Pp. 91–105.

Index

A
B
C 8
D 9
E 0
F 1
G 2
H 3
I 4
J 5